**Sasmita Panda
Dr. Surendra nath Padhi
(Editors)**

Human Diseases

Anchor Academic
Publishing

Panda, Sasmita, Padhi, Surendra nath (Eds.): Human Diseases, Hamburg, Anchor Academic Publishing 2017

Buch-ISBN: 978-3-96067-174-9
PDF-eBook-ISBN: 978-3-96067-674-4
Druck/Herstellung: Anchor Academic Publishing, Hamburg, 2017
Covermotiv: © pixabay.de

Bibliografische Information der Deutschen Nationalbibliothek:
Die Deutsche Nationalbibliothek verzeichnet diese Publikation in der Deutschen Nationalbibliografie; detaillierte bibliografische Daten sind im Internet über http://dnb.d-nb.de abrufbar.

Bibliographical Information of the German National Library:
The German National Library lists this publication in the German National Bibliography. Detailed bibliographic data can be found at: http://dnb.d-nb.de

All rights reserved. This publication may not be reproduced, stored in a retrieval system or transmitted, in any form or by any means, electronic, mechanical, photocopying, recording or otherwise, without the prior permission of the publishers.

Das Werk einschließlich aller seiner Teile ist urheberrechtlich geschützt. Jede Verwertung außerhalb der Grenzen des Urheberrechtsgesetzes ist ohne Zustimmung des Verlages unzulässig und strafbar. Dies gilt insbesondere für Vervielfältigungen, Übersetzungen, Mikroverfilmungen und die Einspeicherung und Bearbeitung in elektronischen Systemen.

Die Wiedergabe von Gebrauchsnamen, Handelsnamen, Warenbezeichnungen usw. in diesem Werk berechtigt auch ohne besondere Kennzeichnung nicht zu der Annahme, dass solche Namen im Sinne der Warenzeichen- und Markenschutz-Gesetzgebung als frei zu betrachten wären und daher von jedermann benutzt werden dürften.

Die Informationen in diesem Werk wurden mit Sorgfalt erarbeitet. Dennoch können Fehler nicht vollständig ausgeschlossen werden und die Diplomica Verlag GmbH, die Autoren oder Übersetzer übernehmen keine juristische Verantwortung oder irgendeine Haftung für evtl. verbliebene fehlerhafte Angaben und deren Folgen.

Alle Rechte vorbehalten

© Anchor Academic Publishing, Imprint der Diplomica Verlag GmbH
Hermannstal 119k, 22119 Hamburg
http://www.diplomica-verlag.de, Hamburg 2017
Printed in Germany

HUMAN DISEASES

EDITED BY

Sasmita Panda
Lecturer in Zoology
Jatni College, Jatni-752050
Odisha, India
E-mail: pandasasmita.2008@gmail.com
&
Dr. S. N. Padhi, Reader Emeritus.
Banki Autonomous College, Banki
Pin- 754008, Odisha, India.
Email- Padhisurendranath5@gmail.com

PREFACE

This book on "Human Diseases" is written with a view to meet curricular requirements of students at undergraduate and post graduate levels of Indian Universities. Disease is illness impairing the normal physiological functioning of living organisms. The causative agents are virus, bacteria, protozoans, helminthes and others. The diseases chosen are measles, smallpox, chickenpox, AIDS, Ebola, Dengue, cholera, leprosy, tuberculosis, typhoid, malaria, amoebiasis, Diarrhoea, Filariasis, Cancer and Jaundice. The text describes causative agents, signs and symptoms, diagnosis, prevention, treatment, epidemiology and history of all the 16 diseases. A table containing these factors have also been depicted for the convenience of readers to have an immediate idea about these human diseases.

We gratefully acknowledge the source of reference at the end of each chapter. The encouragement received by Dr. T. K. Barik, Asst. Prof. Zoology from Prof. Dr. R. P. Das Vice chancellor of Berhampur University, Prof. (Mrs.) S. Patnaik, Prof. (Mrs.) G. Mishra, Dr. P. K. Dixit, Prof. (Mrs.) U. R. Acharya is gratefully acknowledged.

We thank Prof. Dr. A. K. Panda, former principal, Ravenshaw College, Principals of Jatni college, Banki (Autonomous) college, K. B. D. A. V. college, Nirakarpur, Dr. Ajit Rath, Prof. Dr. P. C. Pradhan, Sri G. C. Panda, Sri B. K. Panda, Dr. R. K. Mahapatra and Mrs. Sunita Sadangi for their active participation in preparing this book.

Care has been taken to present the book in error free form but, it is true that "err is human". We apologize for errors if any in the text due to oversight. We will be happy to receive suggestions and mistakes if any, for improvement of the book in the next edition.

Sasmita Panda
S. N. Padhi

ADDRESSES OF CONTRIBUTING AUTHORS

Prof. Dr. Usha Rani Acharya
Rtd Professor of Zoology
Berhampur University
Berhampur-760007
Odisha, India
E-mail: ura_zl@rediffmail.com

Dr. Tapan Kumar Barik
Asst. Professor in Zoology
Post Graduate Department of Zoology
Berhampur University
Berhampur-760007
Odisha, India
E-mail: tkbarik@rediffmail.com

Sasmita Panda
Lecturer in Zoology
Jatni College, Jatni-752050
Odisha, India
E-mail: pandasasmita.2008@gmail.com

Gagan Kumar Panigrahi
Division of Bioscience and Bioinformatics
Myonji University, South Korea
E-mail: gagan.rie@gmail.com

Annapurna Sahoo
Division of Bioscience and Bioinformatics
Myonji University, South Korea
E-mail: annapurna8121990@gmail.com

T. Sarita Achari
Research Scholar
Post Graduate Department of Zoology
Berhampur University
Berhampur-760007
Odisha, India

Simani Mohanty
Research Scholar
Post Graduate Department of Zoology
Berhampur University
Berhampur-760007
Odisha, India

Bibarani Tripathy
Research Scholar
Post Graduate Department of Zoology
Berhampur University
Berhampur-760007
Odisha, India

Surya Narayan Swain
Research Scholar
Post Graduate Department of Zoology
Berhampur University
Berhampur-760007

Mrs. Bijayalaxmi Sahu
Research Scholar
Post Graduate Department of Zoology
Berhampur University
Berhampur-760007
Odisha, India

Deepika Panda
Research Scholar
Post Graduate Department of Zoology
Berhampur University
Berhampur-760007
Odisha, India

Chinmayee Panda
Research Scholar
Post Graduate Department of Zoology
Berhampur University
Berhampur-760007, Odisha, India

Table of Contents

1. MEASLES 1
   - 1.1. INTRODUCTION 1
   - 1.2. SYMPTOMS 1
   - 1.3. COMPLICATIONS 2
   - 1.4. CAUSE 3
   - 1.5. MECHANISM OF INFECTION 4
   - 1.6. DIAGNOSIS 4
   - 1.7. PREVENTION 4
   - 1.8. TREATMENT 4
2. SMALLPOX 7
   - 2.1. INTRODUCTION 7
   - 2.2. SYMPTOMS 7
   - 2.3. COMPLICATIONS 8
   - 2.4. CAUSE 8
   - 2.5. DIAGNOSIS 9
   - 2.6. PREVENTION AND TREATMENT 10
     - 2.6.1. SMALLPOX VACCINE 10
3. CHICKENPOX 12
   - 3.1. INTRODUCTION 12
   - 3.2. CAUSE 12
   - 3.3. SYMPTOMS: 13
     - 3.3.1. Before the rash appears 13
     - 3.3.2. After the rash appears 13
   - 3.4. COMPLICATIONS 13
   - 3.5. DIAGNOSIS 14
   - 3.6. PATHOPHYSIOLOGY 15
     - 3.6.1. Shingles 15
   - 3.7. PREVENTION 15
     - 3.7.1. Hygiene measures 15
     - 3.7.2. Vaccine 16

- 3.8. TREATMENT 16
- 4. AIDS 18
  - 4.1. INTRODUCTION 18
  - 4.2. SYMPTOMS OF HIV 19
  - 4.3. SYMPTOMS OF AIDS 20
  - 4.4. TRANSMISSION 20
  - 4.5. DIAGNOSIS 21
    - 4.5.1. Antibody test 21
    - 4.5.2. Antibody/antigen test 22
    - 4.5.3. Nucleic acid test (NAT) 22
  - 4.6. PREVENTION 22
    - 4.6.1. Sexual contact 22
    - 4.6.2. Pre-exposure 23
    - 4.6.3. Post-exposure 24
  - 4.7. Mother-to-child 24
  - 4.8. VACCINATION 24
  - 4.9. TREATMENT 24
  - 4.10. HIV MEDICATIONS 25
  - 4.11. HIV AND AIDS: CONNECTION 26
- 5. EBOLA 28
  - 5.1. INTRODUCTION 28
  - 5.2. CAUSE AND AGENT 28
  - 5.3. DISEASE AGENT CHARACTERISTICS 28
  - 5.4. GEOGRAPHICAL DISTRIBUTION 29
    - 5.4.1. RESERVOIR 29
  - 5.5. PATHOGENESIS AND TRANSMISSION 30
  - 5.6. SIGNS AND SYMPTOMS 31
  - 5.7. DIAGNOSIS 31
  - 5.8. TREATMENT 32
  - 5.9. PREVENTION 32
- 6. DENGUE 35
  - 6.1. INTRODUCTION 35

6.2. ORIGIN AND HISTORY 35

6.3. DISTRIBUTION 36

6.4. STRUCTURE AND CHARACTERISTICS OF DENGUE VIRUS 36

6.5. REPLICATION OF DENGUE VIRUS 37

6.6. VECTORS OF DENGUE 38

6.7. HOST FACTOR 39

6.8. TRANSMISSION 39

6.9. CLINICAL PRESENTATION 40

6.10. SYMPTOMS 40

6.11. PATHOPHYSIOLOGY 42

6.12. DIAGNOSIS 42

    6.12.1. Clinical laboratory findings 42

    6.12.2. LABORATORY DIAGNOSIS 43

6.13. TREATMENT 44

    6.13.1. Medication 44

    6.13.2. Immunization 44

6.14. PREVENTION 44

7. CHOLERA 46

    7.1. INTRODUCTION 46

    7.2. SYMPTOMS 47

    7.3. CAUSE 47

    7.4. MECHANISM OF TRANSMISSION 48

    7.5. DIAGNOSIS 48

    7.6. TREATMENT 48

    7.7. PREVENTION 49

    7.8. VACCINE 49

8. LEPROSY 51

    8.1. INTRODUCTION 51

    8.2. HISTORY 51

    8.3. GEOGRAPHICAL DISTRIBUTION 51

    8.4. CAUSATIVE AGENT 52

    8.5. TYPES OF LEPROSY 52

- 8.6. TRANSMISSION 53
- 8.7. SYMPTOMS 54
  - 8.7.1. Symptoms caused by damage to the nerves are: 54
  - 8.7.2. Symptoms caused by the disease in the mucous membranes are: 55
- 8.8. COMPLICATIONS 55
- 8.9. DIAGNOSIS 56
  - 8.9.1. SKIN BIOPSY 56
  - 8.9.2. SMEAR TEST 56
  - 8.9.3. SEROLOGY 56
- 8.10. PREVENTION 56
- 8.11. TREATMENT 56

9. TUBERCULOSIS 59
   - 9.1. INTRODUCTION 59
   - 9.2. GEOGRAPHIC DISTRIBUTION 59
   - 9.3. ORIGIN AND HISTORY 59
   - 9.4. CAUSATIVE AGENT AND CNS TUBERCULOSIS 60
   - 9.5. SYMPTOMS 60
   - 9.6. LIFE CYCLE 61
   - 9.7. TRANSMISSION 62
   - 9.8. TREATMENT 62
   - 9.9. TESTING FOR PULMONARY TB 64
     - 9.9.1. SCREENING FOR TB 64
   - 9.10. DIAGNOSIS 64
   - 9.11. DIAGNOSIS OF TB IN INDIA 64
   - 9.12. FLUORESCENT MICROSCOPY 65
   - 9.13. PREVENTION 65

10. TYPHOID 68
    - 10.1. INTRODUCTION 68
    - 10.2. OCCURENCE 68
    - 10.3. EPIDEMIOLOGY 69
    - 10.4. CAUSATIVE AGENT 69
    - 10.5. MORPHOLOGY AND STAINING 70

- 10.6. MULTIPLICATION AND PROPAGATION 70
- 10.7. SYMPTOMS 73
- 10.8. COMPLICATED FEVER 73
- 10.9. CONTAMINATION AND TRANSMISSION 73
- 10.10. DIAGNOSIS 74
- 10.11. TREATMENT 74
- 10.12. SURGERY 75
- 10.13. PREVENTION 75
- 10.14. VACCINATION 75

11. MALARIA 77
- 11.1. INTRODUCTION 77
- 11.2. ORIGIN AND HISTORY 77
- 11.3. GEOGRAPHICAL DISTRIBUTION 77
- 11.4. CAUSATIVE AGENT 78
- 11.5. LIFE CYCLE OF PLASMODIUM 79
- 11.6. VECTORS OF MALARIA 80
- 11.7. TRANSMISSION 81
- 11.8. SYMPTOMS 81
- 11.9. DIAGNOSIS AND TREATMENT 82
- 11.10. PRECAUTIONS 83
- 11.11. PREVENTIVE MEASURES 83
- 11.12. ANTI MALARIAL CAMPAIGN 84

12. AMOEBIASIS 86
- 12.1. INTRODUCTION 86
- 12.2. CAUSATIVE AGENT 86
- 12.3. LIFE CYCLE OF ENTAMOEBA HISTOLYTICA 86
- 12.4. RESERVOIR AND SOURCE 88
- 12.5. TRANSMISSION 88
  - 12.5.1. Faecal-oral route 88
  - 12.5.2. Oral-rectal contact 88
- 12.6. PATHOGENESIS AND PATHOLOGY 89
- 12.7. SYMPTOMS 90

- 12.8. DIAGNOSIS 90
- 12.9. TREATMENT 91
- 12.10. PREVENTION 91

13. DIARRHOEA 93
- 13.1. INTRODUCTION 93
- 13.2. TYPES OF DIARRHOEA 93
  - 13.2.1. Secretory 93
  - 13.2.2. Osmotic 93
  - 13.2.3. Exudative 94
  - 13.2.4. Inflammatory 94
  - 13.2.5. Dysentery 94
- 13.3. DIAGNOSIS 95
- 13.4. CAUSES: 95
- 13.5. PREVENTION 96
- 13.6. TREATMENT 97

14. FILARIASIS 100
- 14.1. INTRODUCTION 100
- 14.2. ORIGIN AND HISTORY 100
- 14.3. CAUSATIVE AGENT 100
- 14.4. MORPHOLOGY OF THE PARASITE 100
- 14.5. LIFE CYCLE OF THE PARASITE 101
- 14.6. DISTRIBUTION 102
- 14.7. CLASSIFICATION OF DISEASE 102
- 14.8. FILARIA VECTORS 102
- 14.9. CAUSE AND MODE OF INFECTION 103
- 14.10. SYMPTOMS 104
- 14.11. BANCROFTIAN FILARIASIS 104
- 14.12. LYMPHATIC FILARIASIS (LF) 105
- 14.13. DIAGNOSIS 105
- 14.14. TREATMENT 106
- 14.15. INDIVIDUAL CHEMOTHERAPY 107
- 14.16. SURGICAL AND SUPPORTIVE TREATMENT 107

- 14.17. PREVENTION ... 107
- 14.18. PERSONAL PRACTICE ... 108
- 15. CANCER ... 109
  - 15.1. INTRODUCTION: ... 109
  - 15.2. WHAT IS CANCER? ... 109
  - 15.3. GENETIC BASIS OF CANCER: ... 110
  - 15.4. ONCOGENES AND SIGNAL TRANSDUCTION: ... 110
  - 15.5. TUMOR SUPPRESSOR GENES: ... 111
  - 15.6. DNA REPAIR GENES: ... 112
  - 15.7. CELL CYCLE: ... 114
  - 15.8. CAUSE OF CANCER: ... 115
  - 15.9. TUMOR BIOLOGY: ... 116
  - 15.10. SIGNS AND SYMPTOMS: ... 118
  - 15.11. DETECTING AND DIAGNOSING CANCER: ... 118
  - 15.12. TREATMENT ... 119
- 16. JAUNDICE ... 124
  - 16.1. INTRODUCTION ... 124
  - 16.2. SIGNS AND SYMPTOMS ... 124
  - 16.3. TYPES OF JAUNDICE ... 125
    - 16.3.1. Pre-hepatic ... 125
    - 16.3.2. Hepatocellular ... 126
  - 16.4. SYMPTOMS ... 126
    - 16.4.1. Post-hepatic ... 127
    - 16.4.2. Neonatal jaundice ... 128
  - 16.5. DIFFERENTIAL DIAGNOSIS ... 128
  - 16.6. PATHOPHYSIOLOGY ... 128
  - 16.7. HEPATIC EVENTS ... 129
  - 16.8. EPIDEMIOLOGY ... 129
    - 16.8.1. DIAGNOSTIC APPROACH ... 129

TABLES

Table 14.1	106
Table 15.1	113
Table 15.2	121
Table 16.1	125
Table 16.2	132

FIGURES

Figure 1.1	3
Figure 6.1	36
Figure 6.2	37
Figure 6.3	38
Figure 6.4	38
Figure 6.5	39
Figure 6.6	41
Figure 8.1	52
Figure 8.2	52
Figure 8.3	54
Figure 8.4	55
Figure 9.1	60
Figure 9.2	61
Figure 10.1	68
Figure 10.2	69
Figure 10.3	71
Figure 10.4	72
Figure 11.1	78
Figure 11.2	79
Figure 11.3	80
Figure 12.1	88
Figure 12.2	89
Figure 14.1	101
Figure 14.2	102
Figure 14.3	103
Figure 14.4	105
Figure 16.1	126

1. MEASLES

Sasmita Panda

1.1. INTRODUCTION

The first systematic description of measles, and its distinction from smallpox and chickenpox, is credited to the Persian physician Rhazes (860–932), who published The Book of Smallpox and Measles. Measles is an endemic disease, meaning it has been continually present in a community, and many people develop resistance. In populations not exposed to measles, exposure to the new disease can be devastating. In 1529, a measles outbreak in Cuba killed two-thirds of those natives who had previously survived smallpox. Two years later, measles was responsible for the deaths of half the population of Honduras, and it had ravaged Mexico, Central America, and the Inca civilization. Measles, or rubeola, is a viral infection of the respiratory system. It is a very contagious disease that can spread through contact with infected mucus and saliva. An infected person can release the infection into the air when they cough or sneeze. The measles virus can live on surfaces for several hours. As the infected particles enter the air and settle on surfaces, anyone within close proximity can become infected. Drinking from an infected person's glass, or sharing eating utensils with an infected person, increases the risk of infection. Measles is a leading cause of death in children. Of the 114,900 global deaths related to measles in 2014, the World Health Organization (WHO) reported that most of the victims were under the age of 5.

1.2. SYMPTOMS

Measles is an infection caused by the rubeola virus. Symptoms will appear about 9 to 11 days after initial infection. Its symptoms always include fever and at least one of the three Cs:

- Cough
- Coryza, or runny nose
- Conjunctivitis
- Symptoms may include:
- Runny nose
- Dry hacking cough
- Conjunctivitis, or swollen eyelids and inflamed eyes
- Watery eyes
- Photophobia, or sensitivity to light
- Sneezing
- A reddish-brown rash

- Koplik's spots or very small greyish-white spots with bluish-white centers in the mouth, insides of cheeks, and throat.
- Generalized body aches

There is often a fever. This can range from mild severe, up to 40.6 degrees Centigrade. It can last several days, and it may fall and then rise again when the rash appears. The reddish-brown rash appears around 3 to 4 days after initial symptoms. This can last for over a week. The rash usually starts behind the ears and spreads over the head and neck. After a couple of days, it spreads to the rest of the body, including the legs. As the spots grow, they often join together. Most childhood rashes are not measles, but a child should see a doctor if:

A parent suspects the child may have measles
Symptoms do not improve, or they get worse
The fever rises to above 38 degrees Centigrade
Other symptoms resolve, but the fever persists

1.3. COMPLICATIONS

The measles vaccine is widely available and is said to have dropped global rates of measles by over 75 percent. Complications from measles are fairly common. Some can be serious. People most at risk are patients with a weak immune system, such as those with HIV, AIDS, leukemia, or a vitamin deficiency, very young children, and adults over the age of 20 years. Older people are more likely to have complications than healthy children over the age of 5 years. Complications can include:

- Diarrhoea
- Vomiting
- Eye infection
- Respiratory tract infections, such as laryngitis and bronchitis
- Difficulty in breathing
- Ear infections, which can lead to permanent hearing loss
- Febrile seizures

Patients with a weakened immune system who have measles are more susceptible to bacterial pneumonia. This can be fatal if not treated. The following less common complications are also possible:

- Hepatitis: Liver complications can occur in adults and in children who are taking some medications.
- Encephalitis: This affects around 1 in every 1,000 patients with measles. It is an inflammation of the brain that can sometimes be fatal. It may occur soon after measles, or several years later.

- Thrombocytopenia, or low platelet count, affects the blood's ability to clot. The patient may bruise easily.
- Squint: Eye nerves and eye muscles may be affected.
- Complications that are very rare but possible include:
- Neuritis, an infection of the optic nerve that can lead to vision loss
- Heart complications
- Subacute sclerosing panencephalitis (SSPE): A brain disease that can affect 2 in every 100,000 people, months or years after measles infection. Convulsions, motor abnormalities, cognitive issues, and death can occur.
- Other nervous system complications include toxic encephalopathy, retrobulbar neuritis, transverse myelitis, and ascending myelitis.
- Measles during pregnancy can lead to miscarriage, early delivery, or low birth weight. A woman who is planning to become pregnant and has not been vaccinated should ask her doctor for advice.

1.4. CAUSE

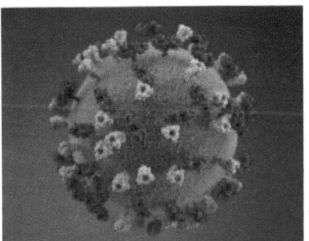

Figure 1.1

An electron micrograph of the measles virus.
Source: www.healthline.com

Measles is caused by infection with the rubeola virus. The virus lives in the mucus of the nose and throat of an infected child or adult. The disease is contagious for 4 days before the rash appears, and it continues to be contagious for about 4 to 5 days after.

Infection spreads through:
- Physical contact with an infected person
- Being near infected people if they cough or sneeze
- Touching a surface that has infected droplets of mucus and then putting fingers into the mouth, or rubbing the nose or eyes. The virus remains active on an object for 2 hours.

1.5. MECHANISM OF INFECTION

As soon as the virus enters the body, it multiplies in the back of the throat, lungs and the lymphatic system. It later infects and replicates in the urinary tract, eyes, blood vessels and central nervous system. The virus takes 1 to 3 weeks to establish itself, but symptoms appear between 9 and 11 days after initial infection. Anyone who has never been infected or vaccinated is likely to become ill if they breathe in infected droplets or are in close physical contact with an infected person. Approximately 90% of people who are not immune will develop measles if they share a house with an infected person.

1.6. DIAGNOSIS

Measles gives people with it a fever. A doctor can normally diagnose measles by looking at the signs and symptoms. A blood test will confirm the presence of the *rubeola virus*. In most countries, measles is a notifiable disease. The doctor has to notify the authorities of any suspected cases. If the patient is a child, the doctor will also notify the school. A child with measles should not return to school until at least 5 days after the rash appears.

Laboratory testing

Alternatively, laboratory diagnosis of measles can be done with confirmation of positive measles IgM antibodies or isolation of measles virus RNA from respiratory specimens. For people unable to have their blood drawn, saliva can be collected for salivary measles-specific IgA testing. Positive contact with other patients known to have measles adds strong epidemiological evidence to the diagnosis. Any contact with an infected person, including semen through sex, saliva, or mucus, can cause infection.

1.7. PREVENTION

Immunizations can help prevent a measles outbreak. The MMR vaccine is a three-in-one vaccination that can protect from the measles, mumps, and rubella (German measles). Children can receive their first MMR vaccination at 12 months, or sooner if traveling internationally, and their second dose between the ages of 4 and 6. Adults who have never received an immunization can request the vaccine from their doctor. If a family member contracts the measles virus, limit interaction with others. This includes staying home from school or work and avoiding social activities. An individual cannot get measles more than once.

1.8. TREATMENT

There is no specific treatment. If there are no complications, the doctor will recommend rest and plenty of fluids to prevent dehydration. Symptoms usually go away within 7 to 10 days.

The following measures may help:

- If the child's temperature is high, they should be kept cool, but not too cold. Tylenol or ibuprofen can help control fever, aches, and pains. Children under 16 years should not take aspirin. A doctor will advise about acetaminophen dosage, as too much can harm the child, especially the liver.
- People should avoid smoking near the child.
- Sunglasses, keeping the lights dim or the room darkened may enhance comfort levels, as measles increases sensitivity to light.
- If there is crustiness around the eyes, gently clean with a warm, damp cloth.
- Cough medicines will not relieve a measles cough. Humidifiers or placing a bowl of water in the room may help. If the child is over 12 months, a glass of warm water with a teaspoon of lemon juice and two teaspoons of honey may help. Do not give honey to infants.
- A fever can lead to dehydration, so the child should drink plenty of fluids.
- A child who is in the contagious stage should stay away from school and avoid close contact with others, especially those who are not immunized or have never had measles.
- Those with a vitamin A deficiency and children under 2 years who have measles may benefit from vitamin A supplements. These can help prevent complications, but they should only be taken with a doctor's agreement.
- Antibiotics will not help against the measles virus, but they may sometimes be prescribed if an additional bacterial infection develops.

REFERENCES

- Baxby D (1997). "Classic Paper: Henry Koplik. The diagnosis of the invasion of measles from a study of the exanthema as it appears on the buccal membrane". Reviews in Medical Virology. 7 (2): 71–4.
- Biesbroeck L, Sidbury R (November 2013). "Viral exanthems: an update". Dermatologic therapy. 26 (6): 433–8.
- GBD 2015 Mortality and Causes of Death, Collaborators. (8 October 2016). "Global, regional, and national life expectancy, all-cause mortality, and cause-specific mortality for 249 causes of death, 1980–2015: a systematic analysis for the Global Burden of Disease Study 2015.". Lancet (London, England). 388 (10053): 1459–1544.
- Kabra, SK; Lodhra, R (14 August 2013). "Antibiotics for preventing complications in children with measles". Cochrane Database of Systematic Reviews. 8: CD001477.

- Ludlow M, McQuaid S, Milner D, de Swart RL, Duprex WP (January 2015). "Pathological consequences of systemic measles virus infection". The Journal of pathology. 235 (2): 253–65.
- Perry RT, Halsey NA (May 1, 2004). "The Clinical Significance of Measles: A Review". The Journal of Infectious Diseases. 189 (S1): S4–16.
- Sabella C (2010). "Measles: Not just a childhood rash". Cleveland Clinic Journal of Medicine. 77 (3): 207–213.
- Sension MG, Quinn TC, Markowitz LE, Linnan MJ, Jones TS, Francis HL, Nzilambi N, Duma MN, Ryder RW (1988). "Measles in hospitalized African children with human immunodeficiency virus". American Journal of Diseases of Children (1960). 142 (12): 1271–2.

2. SMALLPOX

Sasmita Panda

2.1. INTRODUCTION

Smallpox is an extremely contagious and deadly viral disease caused by *Variola virus* for which there is no known cure. The last known case occurred in the United States in 1949 and due to worldwide vaccination programs, this disease has been completely eradicated. Smallpox is also known as variola. Since the time of ancient Egypt, smallpox has proven to be one of the most devastating diseases to humankind. Widespread smallpox epidemics and huge death tolls fill the pages of our history books. The first smallpox vaccine was created in 1758. However, the disease continued to infect and kill people on a widespread basis for another 200 years. The World Health Organization (WHO) implemented a strict vaccination standard in order to slow the infection rate. The last known natural case occurred in 1977 in Somalia. By 1980, the WHO declared that smallpox had been completely eradicated, although Government and health agencies still have stashes of smallpox virus for research purposes. People no longer receive routine smallpox vaccinations. The smallpox vaccine can have potentially fatal side effects, so only the people who are at high risk of exposure get the vaccine.

2.2. SYMPTOMS

Historical accounts show that when someone was infected with the smallpox virus, they had no symptoms for between seven and 17 days. However, once the incubation period (or virus development phase) was over, the following flu-like symptoms occurred:
- High fever
- Chills
- Headache
- Severe back pain
- Abdominal pain
- Vomiting

These symptoms would go away within two to three days. Then the patient would feel better. However, just as the patient started to feel better, a rash would appear. The rash started on the face and then spread to the hands, forearms, and the main part of the body. The person would be highly contagious until the rash disappeared.

Within two days of appearance, the rash would develop into abscesses that filled with fluid and pus. The abscesses would break open and scab over. The scabs would eventually fall off, leaving pit mark scars. Until the scabs fell off, the person remained contagious.

2.3. COMPLICATIONS

Complications of smallpox arise most commonly in the respiratory system and range from simple bronchitis to fatal pneumonia. Respiratory complications tend to develop on about the eighth day of the illness and can be either viral or bacterial in origin. Secondary bacterial infection of the skin is a relatively uncommon complication of smallpox. When this occurs, the fever usually remains elevated.

Other complications include encephalitis (1 in 500 patients), which is more common in adults and may cause temporary disability; permanent pitted scars, most notably on the face; and complications involving the eyes (2 percent of all cases). Pustules can form on the eyelid, conjunctiva, and cornea, leading to complications such as conjunctivitis, keratitis, corneal ulcer, iritis, iridocyclitis, and optic atrophy. Blindness results in approximately 35 percent to 40 percent of eyes affected with keratitis and corneal ulcer. Haemorrhagic smallpox can cause subconjunctival and retinal haemorrhages. In 2 to 5 percent of young children with smallpox, virions reach the joints and bone, causing osteomyelitis variolosa. Lesions are symmetrical, most common in the elbows, tibia, and fibula, and characteristically cause separation of an epiphysis and marked periosteal reactions. Swollen joints limit movement, and arthritis may lead to limb deformities, ankylosis, malformed bones, flail joints, and stubby fingers.

2.4. CAUSE

One of the reasons smallpox was so dangerous and deadly is because it's an airborne disease. Airborne diseases tend to spread fast. Coughing, sneezing, or direct contact with any bodily fluids could spread the smallpox virus. In addition, sharing contaminated clothing or bedding could lead to infection.

TYPES OF SMALLPOX

There were two common and two rare forms of smallpox. The two common forms were known as *variola minor and variola major*. *Variola minor* was a less fatal type of smallpox. The Centers for Disease Control and Prevention (CDC) estimate that only 1 percent of those infected died. However, it was less common than *variola major*. The CDC estimates that 90 percent of smallpox cases were *variola major*. Historically, this type of smallpox killed 30 percent of those infected. The two rare forms of smallpox were known as haemorrhagic and malignant. Both of these rare forms of smallpox carried a very high fatality rate. Hemorrhagic smallpox caused organs to leak blood into the mucous membranes and skin. Malignant smallpox lesions did not develop into pustules or pus-filled bumps on the skin. Instead, they remained soft and flat throughout the entire illness.

TRANSMISSION

Smallpox is an airborne viral disease mainly spread by direct and fairly prolonged face-to-face contact between people. Smallpox patients became contagious once the first sores appeared in their mouth and throat (early rash stage). They spread the virus when they coughed or sneezed and droplets from their nose or mouth spread to other people. They remained contagious until their last smallpox scab fell off. These scabs and the fluid found in the patient's sores also contained the *variola virus*. The virus can spread through these materials or through the objects contaminated by them, such as bedding or clothing. People who cared for smallpox patients and washed their bedding or clothing had to wear gloves and take care to not get infected. Rarely, smallpox has spread through the air in enclosed settings, such as a building (airborne route). Smallpox can be spread by humans only. Scientists have no evidence that smallpox can be spread by insects or animals.

2.5. DIAGNOSIS

Smallpox is defined as an illness with acute onset of fever equal to or greater than 38.3 °C (101 °F) followed by a rash characterized by firm, deep seated vesicles or pustules in the same stage of development without other apparent cause. If a clinical case is observed, smallpox is confirmed using laboratory tests.

Microscopically, pox viruses produce characteristic cytoplasmic inclusions, the most important of which are known as Guarnieri bodies, and are the sites of viral replication. Guarnieri bodies are readily identified in skin biopsies stained with hematoxylin and eosin, and appear as pink blobs. They are found in virtually all pox virus infections but the absence of Guarnieri bodies cannot be used to rule out smallpox. The diagnosis of an orthopox virus infection can also be made rapidly by electron microscopic examination of pustular fluid or scabs. However, all orthopox viruses exhibit identical brick-shaped virions by electron microscopy. However, if particles with the characteristic morphology of herpes viruses are seen this will eliminate smallpox and other orthopox virus infections.

Definitive laboratory identification of *variola virus* involves growing the virus on chorioallantoic membrane (part of a chicken embryo) and examining the resulting pock lesions under defined temperature conditions. Strains may be characterized by polymerase chain reaction (PCR) and restriction fragment length polymorphism (RFLP) analysis. Serologic tests and enzyme linked immunosorbent assays (ELISA), which measure variola virus-specific immunoglobulin and antigen have also been developed to assist in the diagnosis of infection.

Chickenpox was commonly confused with smallpox in the immediate post-eradication era. Chickenpox and smallpox can be distinguished by several methods. Unlike smallpox, chickenpox does not usually affect the palms and soles. Additionally, chickenpox pustules are of varying size due to variations in the timing of pustule eruption: smallpox pustules are all very nearly the same size since the viral effect progresses more uniformly. A variety of laboratory methods are available for detecting chickenpox in evaluation of suspected smallpox cases.

2.6. PREVENTION AND TREATMENT

There is no proven treatment for smallpox disease, but some antiviral drugs may help treat it or prevent it from getting worse. There also is a vaccine to protect people from smallpox. If there were a smallpox outbreak, health officials would use the smallpox vaccine to control it.

2.6.1. SMALLPOX VACCINE

Before contact with the virus, the vaccine can protect you from getting sick.

Within 3 days of being exposed to the virus, the vaccine might protect you from getting the disease. If you still get the disease, you might get much less sick than an unvaccinated person would.

Within 4 to 7 days of being exposed to the virus, the vaccine likely gives you some protection from the disease. If you still get the disease, you might not get as sick as an unvaccinated person would.

Once an individual have developed the smallpox rash, the vaccine will not protect him. Currently, the smallpox vaccine is not available to the general public because smallpox has been eradicated, and the virus no longer exists in nature.

REFERENCES

- Barquet N, Domingo P (15 October 1997). "Smallpox: the triumph over the most terrible of the ministers of death". Annals of Internal Medicine. 127 (8 Pt 1): 635–42.
- Bray M, Roy CJ (2004). "Antiviral prophylaxis of smallpox". J. Antimicrob. Chemother. 54 (1): 1–5.
- Dubochet J, Adrian M, Richter K, Garces J, Wittek R (1994). "Structure of intracellular mature vaccinia virus observed by cryoelectron microscopy". J. Virol. 68 (3): 1935–41.
- Duggan AT, Perdomo MF, Piombino-Mascali D, Marciniak S, Poinar D, Emery MV, Buchmann JP, Duchêne S, Jankauskas R, Humphreys M, Golding GB, Southon J, Devault A, Rouillard JM, Sahl JW, Dutour O, Hedman K, Sajantila A, Smith GL, Holmes EC, Poinar HN. "17th Century Variola Virus Reveals the Recent History of Smallpox". Curr Biol. 26: 3407–3412.
- Elizabeth A. Fenn. "Biological Warfare in Eighteenth-Century North America: Beyond Jeffery Amherst", The Journal of American History, Vol. 86, No. 4 (March 2000), pp. 1552–1580.
- Esposito JJ, Sammons SA, Frace AM, Olsen-Rasmussen M, et al. (August 2006). "Genome sequence diversity and clues to the evolution of *variola* (smallpox) virus". Science. 313 (5788): 807–12.

- Fenner, Frank (1988). "Development of the Global Smallpox Eradication Programme" (PDF). Smallpox and Its Eradication (History of International Public Health, No. 6). Geneva: World Health Organization. pp. 366–418.
- Henderson DA, Inglesby TV, Bartlett JG, Ascher MS, Eitzen E, Jahrling PB, Hauer J, Layton M, McDade J, Osterholm MT, O'Toole T, Parker G, Perl T, Russell PK, Tonat K (1999). "Smallpox as a biological weapon: medical and public health management. Working Group on Civilian Biodefense". JAMA. 281 (22): 2127–37.
- Hughes AL, Irausquin S, Friedman R (2010). "The evolutionary biology of pox viruses". Infection, Genetics and Evolution. 10 (1): 50–59.
- Jezek Z, Hardjotanojo W, Rangaraj AG (1981). "Facial scarring after varicella. A comparison between *variola major* and *variola minor*". Am. J. Epidemiol. 114 (6): 798–803.
- Rodrigues BA (1975). "Smallpox eradication in the Americas". Bull Pan Am Health Organ. 9 (1): 53–68.
- Ryan KJ, Ray CG, eds. (2004). Sherris Medical Microbiology (4th ed.). McGraw Hill. pp. 525–8. ISBN 0-8385-8529-9.
- Shchelkunov SN (December 2011). "Emergence and reemergence of smallpox: the need for development of a new generation smallpox vaccine". Vaccine. 29 Suppl 4: D49–53.
- Tognotti E. (June 2010). "The eradication of smallpox, a success story for modern medicine and public health: What lessons for the future?" (PDF). J Infect Dev Ctries. 4 (5): 264–266.
- Wujastyk, Dominik. (1995). "Medicine in India," in Oriental Medicine: An Illustrated Guide to the Asian Arts of Healing, 19–38. Edited by Serindia Publications. London: Serindia Publications. ISBN 0-906026-36-9. p. 29.

3. CHICKENPOX

Sasmita Panda

3.1. INTRODUCTION

Chickenpox is an airborne disease which spreads easily through the coughs and sneezes of an infected person. It may be spread from one to two days before the rash appears until all lesions have crusted over. It may also spread through contact with the blisters. Those with shingles may spread chickenpox to those who are not immune through contact with the blisters. The disease can usually be diagnosed based on the presenting symptom; however, in unusual cases it may be confirmed by polymerase chain reaction (PCR) testing of the blister fluid or scabs. Testing for antibodies may be done to determine if a person is or is not immune. People usually only get the disease once. Although reinfections by the virus occur, these reinfections usually do not cause any symptoms.

The varicella vaccine has resulted in a decrease in the number of cases and complications from the disease. It protects about 70 to 90 percent of people from disease with a greater benefit for severe disease. Routine immunization of children is recommended in many countries. Immunization within three days of exposure may improve outcomes in children. Treatment of those infected may include calamine lotion to help with itching, keeping the fingernails short to decrease injury from scratching, and the use of paracetamol (acetaminophen) to help with fevers. For those at increased risk of complications antiviral medication such as aciclovir are recommended. Chickenpox occurs in all parts of the world. As of 2013 140 million cases of chickenpox and herpes zoster occurred. Before routine immunization the number of cases occurring each year was similar to the number of people born. Since immunization the number of infections in the United States has decreased nearly 90%. In 2015 chickenpox resulted in 6,400 deaths globally – down from 8,900 in 1990. Death occurs in about 1 per 60,000 cases. Chickenpox was not separated from smallpox until the late 19th century. In 1888 its connection to shingles was determined. The first documented use of the term chicken pox was in 1658. Various explanations have been suggested for the use of "chicken" in the name, one being the relative mildness of the disease. Humans are the only known animal that the disease affects naturally. However, chickenpox has been caused in other animals, such as primates, including chimpanzees and gorillas.

3.2. CAUSE

Chickenpox, also known as varicella, is a highly contagious disease caused by the initial infection with *varicella zoster virus* (VZV). The disease results in a characteristic skin rash that forms small, itchy blisters, which eventually scab over. It usually starts on the chest, back, and face then spreads to the rest of the body. Other symptoms may include fever, feeling tired, and headaches. Symptoms usually last for five to ten days. Complications may occasionally include pneumonia, inflammation of the brain, or bacterial infections of the skin among others. The disease is often more severe in adults than children. Symptoms begin ten to twenty-one days after exposure to the virus.

3.3. SYMPTOMS

The following are common chickenpox symptoms:

3.3.1. Before the rash appears

- A general feeling of being unwell (malaise).
- Fever (usually worse in adults than children)
- Aching muscles.
- Loss of appetite.
- Sometimes there may be a feeling of nausea.

3.3.2. After the rash appears

- Rash - its severity varies considerably. Some patients may have just a few spots, while others are covered all over the body.
- Spots - the spots, which develop in clusters, generally appear on the face, limbs, chest, and stomach.
- Blisters - initially, there are small red spots that itch a lot. They then develop into spots with blisters on top - these can become very itchy.
- Clouding - within about 48 hours, the blisters cloud over and start drying out (a crust develops).
- Healing - within about 10 days, the crusts fall off on their own.
- During the whole cycle, new waves of spots can appear - in such cases, the patient might have different clusters of spots at varying stages of itchiness, dryness, and crustiness.

3.4. COMPLICATIONS

Most healthy individuals make a full recovery, as is the case with a cold or flu - just by resting and drinking plenty of fluids. A small percentage of patients have more severe symptoms such as,

the skin around the spots or blisters becomes painful and red

there are breathing difficulties

PREGNANCY AND NEONATES

During pregnancy the dangers to the fetus associated with a primary VZV infection are greater in the first six months. In the third trimester, the mother is more likely to have severe symptoms. For pregnant women, antibodies produced as a result of immunization or previous infection are transferred via the placenta to the foetus. Women who are immune to chickenpox cannot become infected and do not need to be concerned about it for themselves or their infant during pregnancy.

Varicella infection in pregnant women could lead to spread via the placenta and infection of the foetus. If infection occurs during the first 28 weeks of gestation, this can lead to foetal varicella syndrome (also known as congenital varicella syndrome). Effects on the foetus can range in severity from underdeveloped toes and fingers to severe anal and bladder malformation. Possible problems include:

Damage to brain: encephalitis, microcephaly, hydrocephaly, aplasia of brain

Damage to the eye: optic stalk, optic cup, and lens vesicles, microphthalmia, cataracts, chorioretinitis, optic atrophy

Other neurological disorder: damage to cervical and lumbosacral spinal cord, motor/sensory deficits, absent deep tendon reflexes, anisocoria/Horner's syndrome

Damage to body: hypoplasia of upper/lower extremities, anal and bladder sphincter dysfunction

Skin disorders: (cicatricial) skin lesions, hypopigmentation

Infection late in gestation or immediately following birth is referred to as "neonatal varicella". Maternal infection is associated with premature delivery. The risk of the baby developing the disease is greatest following exposure to infection in the period 7 days before delivery and up to 8 days following the birth. The baby may also be exposed to the virus via infectious siblings or other contacts, but this is of less concern if the mother is immune. Newborns who develop symptoms are at a high risk of pneumonia and other serious complications of the disease.

3.5. DIAGNOSIS

The diagnosis of chickenpox is primarily based on the signs and symptoms, with typical early symptoms followed by a characteristic rash. Confirmation of the diagnosis is by examination of the fluid within the vesicles of the rash, or by testing blood for evidence of an acute immunologic response. Vesicular fluid can be examined with a Tzanck smear, or better by testing for direct fluorescent antibody. The fluid can also be "cultured", whereby attempts are made to grow the virus from a fluid sample. Blood tests can be used to identify a response to acute infection (IgM) or previous infection and subsequent immunity (IgG). Prenatal diagnosis of fetal varicella infection can be performed using ultrasound, though a delay of 5 weeks following primary maternal infection is advised. A PCR (DNA) test of the mother's amniotic fluid can also be performed, though the risk of spontaneous abortion due to the amniocentesis procedure is higher than the risk of the baby's developing fetal varicella syndrome.

3.6. PATHOPHYSIOLOGY

Exposure to VZV in a healthy child initiates the production of host immunoglobulin G (IgG), immunoglobulin M (IgM), and immunoglobulin A (IgA) antibodies; IgG antibodies persist for life and confer immunity. Cell-mediated immune responses are also important in limiting the scope and the duration of primary varicella infection. After primary infection, VZV is hypothesized to spread from mucosal and epidermal lesions to local sensory nerves. VZV then remains latent in the dorsal ganglion cells of the sensory nerves. Reactivation of VZV results in the clinically distinct syndrome of herpes zoster (i.e., shingles), postherpetic neuralgia, and sometimes Ramsay Hunt syndrome type II. Varicella zoster can affect the arteries in the neck and head, producing stroke, either during childhood, or after a latency period of many years.

3.6.1. Shingles

Chickenpox and shingles are caused by the same virus. Shingles occurs when the *varicella zoster virus* from a previous case of chicken pox becomes active again. A person who has never had chickenpox or was never vaccinated can catch chickenpox from someone with shingles. However, you cannot get shingles from somebody with chickenpox.

Complications of shingles can include:

Postherpetic neuralgia - pain from shingles lasts long after the blisters have gone.

Vision loss - from eye infections caused shingles in the area.

Neurological problems - due to inflammation in the brain.

Skin infections - more likely if blisters are not treated correctly.

Shingles affects one in five adults infected with chickenpox as children, especially those who are immune-suppressed, particularly from cancer, HIV, or other conditions. Stress can bring on shingles as well, although scientists are still researching the connection. Shingles are most commonly found in adults over the age of 60 who were diagnosed with chickenpox when they were under the age of 1.

3.7. PREVENTION

3.7.1. Hygiene measures

The spread of chickenpox can be prevented by isolating affected individuals. Contagion is by exposure to respiratory droplets, or direct contact with lesions, within a period lasting from three days before the onset of the rash, to four days after the onset of the rash. The chickenpox virus is susceptible to disinfectants, notably chlorine bleach (i.e., sodium hypochlorite). Like all enveloped viruses, it is sensitive to desiccation, heat and detergents.

3.7.2. Vaccine

The chickenpox vaccine prevents chickenpox in 90 percent of children who receive it. The shot should be given when a child is between 12 and 15 months of age. A booster is given between 4 and 6 years of age. Older children and adults who have not been vaccinated or exposed may receive catch-up doses of the vaccine. As chickenpox tends to be more severe in older patients, parents who did not previously vaccinate may opt to have the shots given later.

People who are unable to receive the vaccine can try to avoid the virus by limiting contact with infected people. This can be difficult, as chickenpox can't be identified by blisters until it has been contagious for days.

3.8. TREATMENT

Chickenpox generally resolves within a week or two without treatment - there is no medication or treatment to "cure" it. A vaccine is available for varicella. For children, 2 doses of the varicella vaccine (one given at 12-15 months and one given at age 4-6) are 90 percent effective at preventing chickenpox. A doctor may prescribe or advise on how to reduce symptoms of itchiness and discomfort, and also on how to prevent the infection from spreading to other people.

Pain or fever - Tylenol (acetaminophen) may help with symptoms of high temperature and/or pain. It is important to follow the instructions provided by the manufacturer. Aspirin containing products should NOT be used for chickenpox as this can lead to complications. Pregnant women can take acetaminophen (Tylenol) at any time during their pregnancy.

Avoiding dehydration - the patient should drink plenty of fluids, preferably water, to prevent dehydration. Some doctors recommend sugar-free popsicles or Pedialyte for children who are not drinking enough.

Mouth soreness - sugar-free popsicles help ease symptoms of soreness if there are spots in the mouth. Salty or spicy foods should be avoided. If chewing is painful, soup might be a good option; make sure it is not too hot.

Itchiness - although itchiness can become severe and the urge to scratch may seem impossible to control, it is important to keep scratching down to a minimum to prevent future scarring of the skin. The following may help a child with chickenpox:

- Keep fingernails as short as possible.
- Keep fingernails clean at all times.
- Place mittens or even socks over the child's hands when they go to sleep so that any attempt at scratching during the night does not cut the skin.
- Calamine lotions or oatmeal baths may help reduce symptoms of itching.
- Make sure the patient wears only loose clothing.

Antiviral medication – It may be prescribed to pregnant females, adults if a diagnosis is made early enough, newborns, and patients with weakened immune systems. An example of such a drug is Acyclovir. This medication works best if it is given within 24 hours of developing symptoms. Acyclovir reduces the severity of symptoms but does not cure the disease.

REFERENCES

- Askalan R, Laughlin S, Mayank S, Chan A, MacGregor D, Andrew M, Curtis R, Meaney B, deVeber G (June 2001). "Chickenpox and stroke in childhood: a study of frequency and causation". Stroke. 32 (6): 1257–62.
- Boussault P, Boralevi F, Labbe L, Sarlangue J, Taïeb A, Leaute-Labreze C (2007). "Chronic varicella-zoster skin infection complicating the congenital varicella syndrome". Pediatr Dermatol. 24 (4): 429–32.
- Breuer J (2010). "VZV molecular epidemiology". Current Topics in Microbiology and Immunology. 342: 15–42.
- Flatt, A; Breuer, J (September 2012). "Varicella vaccines.". British medical bulletin. 103 (1): 115–27.
- Heuchan AM, Isaacs D (19 March 2001). "The management of varicella-zoster virus exposure and infection in pregnancy and the newborn period. Australasian Subgroup in Paediatric Infectious Diseases of the Australasian Society for Infectious Diseases.". The Medical journal of Australia. 174 (6): 288–92.
- Kanbayashi Y, Onishi K, Fukazawa K, Okamoto K, Ueno H, Takagi T, Hosokawa T (2012). "Predictive Factors for Postherpetic Neuralgia Using Ordered Logistic Regression Analysis". The Clinical Journal of Pain. 28 (8): 712–714.
- Macartney, K; Heywood, A; McIntyre, P (23 June 2014). "Vaccines for post-exposure prophylaxis against varicella (chickenpox) in children and adults.". The Cochrane database of systematic reviews. 6: CD001833.
- Matsuo T, Koyama M, Matsuo N (July 1990). "Acute retinal necrosis as a novel complication of chickenpox in adults". Br J Ophthalmol. 74 (7): 443–4.
- Nagel MA, Cohrs RJ, Mahalingam R, Wellish MC, Forghani B, Schiller A, Safdieh JE, Kamenkovich E, Ostrow LW, Levy M, Greenberg B, Russman AN, Katzan I, Gardner CJ, Häusler M, Nau R, Saraya T, Wada H, Goto H, de Martino M, Ueno M, Brown WD, Terborg C, Gilden DH (March 2008). "The varicella zoster virus vasculopathies: clinical, CSF, imaging, and virologic features.". Neurology. 70: 853–60.
- Pino Rivero V, González Palomino A, Pantoja Hernández CG, Mora Santos ME, Trinidad Ramos G, Blasco Huelva A (2006). "Ramsay-Hunt syndrome associated to unilateral recurrential paralysis". Anales otorrinolaringologicos ibero-americanos. 33 (5): 489–494.

4. AIDS

G. K. Panigrahi

4.1. INTRODUCTION

HIV is a virus that enters your body and begins to destroy T cells. T cells are needed in order to fight infections. HIV spreads through bodily fluids that include: blood, semen, vaginal and rectal fluids, breast milk.

The first few weeks after infection is called the acute infection stage. During this time the virus rapidly reproduces. Your immune system responds by producing HIV antibodies. Many people experience temporary flu-like symptoms during this stage. Even without symptoms, HIV is highly contagious during this time. After the first month or so, HIV enters the clinical latency stage. This stage can last from a few years to a few decades. Progression can be slowed with antiretroviral therapy. Some people have symptoms. Many people do not, but it's still contagious. As the virus progresses, you're left with fewer T cells. This makes you more susceptible to disease, infection, and infection-related cancers. HIV is a lifetime condition with no cure. Medical care, including antiretroviral therapy, can help manage HIV and prevent AIDS. Without treatment, HIV is likely to advance to AIDS. At that point, the immune system is too weak to fight off life-threatening disease and infection. Untreated, life expectancy with AIDS is about three years. It is estimated that 1.1 million Americans are currently living with HIV. And one in five don't know it.

AIDS is a disease caused by HIV. It's the most advanced stage of HIV. But just because you have HIV doesn't mean you'll develop AIDS. HIV destroys T cells called CD4 cells. These cells help your immune system fight infections. Healthy adults generally have a CD4 count of 800 to 1,000 per cubic millimeter. If you have HIV and your CD4 count falls below 200 per cubic millimeter, you will be diagnosed with AIDS. One can also be diagnosed with AIDS if one have HIV and develop an opportunistic infection that is rare in people who don't have HIV. AIDS weakens your immune system to the point where it can no longer fight off most diseases and infections.

In addition, women with HIV are at increased risk of:
- vaginal yeast infections and other vaginal infections, including bacterial vaginosis
- sexually transmitted diseases (STDs) such as gonorrhea, chlamydia, and trichomoniasis
- pelvic inflammatory disease (PID)
- infection of the reproductive organs and menstrual cycle changes
- human *papilloma virus* (HPV), which can cause genital warts and lead to cervical cancer

- Another gender difference is that women are less likely than men to notice small spots or other changes to their genitals.
- HIV can be transmitted to your baby during pregnancy. The virus can also be passed to your baby through breast milk. If your doctor knows you have HIV, treatment can lower the risk of passing the virus on to your child to less than 2 percent.

4.2. SYMPTOMS OF HIV

Some people infected with HIV are asymptomatic at first. Most people experience symptoms in the first month or two after becoming infected. That's because your immune system is reacting to the virus as it rapidly reproduces.

This early stage is called acute stage. Symptoms are similar to those of the flu and may last anywhere from a few days to several weeks. These include:
- Fever
- Swollen lymph glands
- General aches and pains

During the first few months of infection, an HIV test may provide a false-negative result. This is because it takes time for the immune system to build up enough antibodies to be detected in a blood test. But the virus is active and highly contagious during this time.

The clinical latent infection, or chronic stage of HIV, can last from a few years to a few decades. During this time the virus is still reproducing, but at lower levels. Some people have few, if any, symptoms. Others may have many symptoms. As the disease progresses, other symptoms may include:
- Swollen lymph nodes
- Recurrent fevers
- Fatigue
- Aches and pains
- Nausea, vomiting
- Diarrhoea
- Weight loss
- Skin rashes
- Oral yeast infections or other infections
- Shingles
- Symptoms may come and go or progress rapidly. Even if no symptoms, an individual can still transmit the virus to others.

4.3. SYMPTOMS OF AIDS

With the use of antiretroviral therapy, chronic HIV can last several decades. Without treatment, HIV can be expected to progress to AIDS sooner. By that time, the immune system is quite damaged and has a hard time fighting off infection and disease. Symptoms of AIDS can include:

- Recurrent fever
- Chronic swollen lymph glands, especially of the armpits, neck, and groin
- Chronic fatigue
- Night sweats
- Dark splotches under the skin or inside the mouth, nose, or eyelids
- Sores, spots, or lesions of the mouth and tongue, genitals, or anus
- Bumps, lesions, or rashes of the skin
- Recurrent or chronic diarrhoea
- Rapid weight loss
- Neurologic problems such as difficulty concentrating, memory loss, and confusion
- Anxiety and depression
- An individual with a weakened immune system has an increased risk of pneumonia and other opportunistic infections. Other potential complications of AIDS include:
- Candidiasis.
- Tuberculosis.
- *Cytomegalo virus* (CMV), a type of *herpes virus.*
- Cryptococcal meningitis.
- Toxoplasmosis, and infection caused by a parasite.
- Cryptosporidiosis, an infection caused by an intestinal parasite.
- Cancer, including Kaposi's sarcoma (KS) and lymphoma and
- Kidney disease.
- Antiviral medications can help control the virus. Treatment for other infections and complications of AIDS must be tailored to an individual needs.

4.4. TRANSMISSION

HIV does not play favorites. Anyone can become infected. The virus is transmitted in bodily fluids that include:

- Blood
- Semen
- Vaginal and rectal fluids
- Breast milk

Some of the ways HIV is spread from person to person include:
- Having unprotected sex with an infected person. This is the most common route of transmission
- Sharing needles, syringes, and other items for injection drug use with an infected person
- Passing it on to an unborn child if the mother is HIV-positive
- Passing it on to a baby through breast milk if the mother is HIV-positive
- Being exposed to infected fluids, usually through needle sticks.
- Having a blood transfusion or organ and tissue transplant can also transmit the virus. But rigorous testing for HIV in blood ensures that this is very rare in the United States.
- It's theoretically possible, but considered extremely rare, for HIV to spread via:
- Oral sex
- Being bitten by an infected person
- Contact between broken skin, wounds, or mucous membranes and HIV-infected blood or fluids

HIV does NOT spread through:
- Skin-to-skin contact
- Hugging, shaking hands, or kissing
- Air or water
- Eating or drinking items, including drinking fountains
- Saliva, tears, or sweat (unless mixed with blood from an infected person)
- Sharing a toilet, towels, or bedding
- Mosquitoes or other insects

4.5. DIAGNOSIS

4.5.1. Antibody test

Between 21 and 84 days after infection, about 97 percent of people will develop detectable HIV antibodies, which can be found in the blood or saliva.

There's no preparation necessary for blood tests or mouth swabs. Some tests provide results in 30 minutes or less and can be performed in a doctor's office or clinic. There are also home test kits available:
- Oral Quick HIV Test: An oral swab provides results in as little as 20 minutes.
- Home Access HIV-1 Test System: After pricking your finger, you send a blood sample to a licensed laboratory. You can remain anonymous and call for results the next business day.

- If you think you've recently been exposed to HIV, but tested negative, repeat the test in three months. If you have a positive result, follow up with your doctor to confirm.

4.5.2. Antibody/antigen test

An antigen is part of the virus that activates the immune system. It takes from 13 to 42 days for antibodies and antigens to be detectable.

4.5.3. Nucleic acid test (NAT)

This expensive test isn't used for general screening. It's for people who have early symptoms of HIV or recently had a high-risk exposure. This test doesn't look for antibodies, but for the virus itself. It takes from seven to 28 days for HIV to be detectable in the blood. This test is usually accompanied by an antibody test.

About 90 percent of people with HIV experience changes to the skin. Rash is often one of the first symptoms of HIV infection. Generally, an HIV rash appears as a flat red area with small bumps. HIV makes you more susceptible to skin problems because the virus destroys immune system cells that fight infection. Co-infections that can cause rash include:
- Molluscum contagiosum
- Herpes simplex
- Shingles

The appearance of the rash, how long it lasts, and how it can be treated depend on the cause.

Some medicines used to treat HIV or other infections can cause a rash. It usually appears within a week or two of starting on a new medication. Sometimes the rash will clear up on its own. If it doesn't, you may need to switch medicines. Rash due to an allergic reaction to medicine can be serious. Other symptoms of an allergic reaction include trouble breathing or swallowing, dizziness, and fever. Stevens-Johnson syndrome (SJS) is a rare allergic reaction to HIV medication. Symptoms include fever and swelling of the face and tongue. Rash, which can involve the skin and mucous membrane, appears and spreads quickly.

When 30 percent of the skin is affected it's called toxic epidermal necrolysis, which is a life-threatening condition.

4.6. PREVENTION

4.6.1. Sexual contact

Consistent condom use reduces the risk of HIV transmission by approximately 80% over the long term. When condoms are used consistently by a couple in which one person is infected, the rate of HIV infection is less than 1% per year. There is some evidence to suggest that female condoms may provide an equivalent level of protection. Application of a vaginal gel containing tenofovir (a reverse transcriptase inhibitor) immediately before sex seems to reduce infection rates by approximately 40% among African women. By contrast, use of the spermicide nonoxynol-9 may increase the risk of transmission due to its tendency to cause vaginal and rectal irritation.

Circumcision in Sub-Saharan Africa "reduces the acquisition of HIV by heterosexual men by between 38% and 66% over 24 months". Due to these studies, both the World Health Organization and UNAIDS recommended male circumcision as a method of preventing female-to-male HIV transmission in 2007 in areas with a high rates of HIV. However, whether it protects against male-to-female transmission is disputed, and whether it is of benefit in developed countries and among men who have sex with men is undetermined. The International Antiviral Society, however, does recommend for all sexually active heterosexual males and that it be discussed as an option with men who have sex with men. Some experts fear that a lower perception of vulnerability among circumcised men may cause more sexual risk-taking behavior, thus negating its preventive effects.

Programs encouraging sexual abstinence do not appear to affect subsequent HIV risk. Evidence of any benefit from peer education is equally poor. Comprehensive sexual education provided at school may decrease high risk behavior. A substantial minority of young people continues to engage in high-risk practices despite knowing about HIV/AIDS, underestimating their own risk of becoming infected with HIV. Voluntary counseling and testing people for HIV does not affect risky behavior in those who test negative but does increase condom use in those who test positive. It is not known whether treating other sexually transmitted infections is effective in preventing HIV.

4.6.2. Pre-exposure

Antiretroviral treatment among people with HIV whose CD4 count ≤ 550 cells/µL is a very effective way to prevent HIV infection of their partner (a strategy known as treatment as prevention, or TASP). TASP is associated with a 10 to 20 fold reduction in transmission risk. Pre-exposure prophylaxis (PrEP) with a daily dose of the medications tenofovir, with or without emtricitabine, is effective in a number of groups including men who have sex with men, couples where one is HIV positive, and young heterosexuals in Africa. It may also be effective in intravenous drug users with a study finding a decrease in risk of 0.7 to 0.4 per 100 person years.

Universal precautions within the health care environment are believed to be effective in decreasing the risk of HIV. Intravenous drug use is an important risk factor and harm reduction strategies such as needle-exchange programs and opioid substitution therapy appear effective in decreasing this risk.

4.6.3. Post-exposure

A course of antiretrovirals administered within 48 to 72 hours after exposure to HIV-positive blood or genital secretions is referred to as post-exposure prophylaxis (PEP). The use of the single agent zidovudine reduces the risk of a HIV infection five-fold following a needle-stick injury. As of 2013, the prevention regimen recommended in the United States consists of three medications—tenofovir, emtricitabine and raltegravir—as this may reduce the risk further.

PEP treatment is recommended after a sexual assault when the perpetrator is known to be HIV positive, but is controversial when their HIV status is unknown. The duration of treatment is usually four weeks and is frequently associated with adverse effects—where zidovudine is used, about 70% of cases result in adverse effects such as nausea (24%), fatigue (22%), emotional distress (13%) and headaches (9%).

4.7. Mother-to-child

Programs to prevent the vertical transmission of HIV (from mothers to children) can reduce rates of transmission by 92–99%. This primarily involves the use of a combination of antiviral medications during pregnancy and after birth in the infant and potentially includes bottle feeding rather than breast feeding. If replacement feeding is acceptable, feasible, affordable, sustainable, and safe, mothers should avoid breast feeding their infants; however exclusive breast feeding is recommended during the first months of life if this is not the case. If exclusive breast feeding is carried out, the provision of extended antiretroviral prophylaxis to the infant decreases the risk of transmission. In 2015, Cuba became the first country in the world to eradicate mother-to-child transmission of HIV.

4.8. VACCINATION

Currently, there is no licensed vaccine for HIV or AIDS. The most effective vaccine trial to date, RV 144, was published in 2009 and found a partial reduction in the risk of transmission of roughly 30%, stimulating some hope in the research community of developing a truly effective vaccine. Further trials of the RV 144 vaccine are ongoing.

4.9. TREATMENT

Treatment should begin as soon as possible after a diagnosis of HIV. The main treatment for HIV is antiretroviral therapy (ART), a combination of daily medications that stop the virus from reproducing. This helps protect your CD4 cells, keeping your immune system strong enough to fight off disease. ART helps keep HIV from progressing to AIDS. It also helps reduce the risk of transmission. There are more than 25 medications in six drug classes approved to treat HIV. The U.S. Department of Health and Human Services (HHS) recommends a starting regimen of three HIV medicines from at least two drug classes. Doctor will help an individual choose a regimen based on your overall health and personal circumstances. These medications must be taken consistently and exactly as prescribed. Failure to adhere to therapy guidelines can jeopardize your health. Side effects vary and may include headache and dizziness. Serious side effects include swelling of the mouth and tongue and liver damage. Some people eventually develop drug-resistant strains of HIV. If you have serious side effects, your medications can be adjusted.

Doctor may also recommend vaccinations for the following conditions:
- Hepatitis B
- Influenza
- Pneumonia

Treatment for individual symptoms can be addressed as they arise. To strengthen overall health, maintain a healthy diet, exercise regularly, and get enough sleep.

4.10. HIV MEDICATIONS

There are at least 25 medications approved to treat HIV. They work to prevent HIV from reproducing and destroying CD4 cells, which help your immune system fight infection. This also helps reduce the risk of transmitting the virus.

These antiretroviral medications are grouped into six classes:
- Non-nucleoside reverse transcriptase inhibitors (NNRTIs)
- Nucleoside reverse transcriptase inhibitors (NRTIs)
- Protease inhibitors
- Fusion inhibitors
- CCR5 antagonists, also known as entry inhibitors
- Integrase strand transfer inhibitors
- Treatment should begin as quickly as possible and usually starts with a regimen of three drugs from at least two classes. Your health will help determine your best options.
- Antiretroviral medications must be taken exactly as prescribed to be effective. Some are available in combination pills so they're easier to take.
- Side effects differ from person to person. The most common are dizziness and headache. Serious side effects may include swelling of the mouth and tongue and liver damage. Drug interactions and drug resistance are also possible.
- Blood testing will help determine if the regimen is working to keep your viral count down and your CD4 count up.
- Costs vary according to where you live and type of insurance coverage. Some pharmaceutical companies have assistance programs to lower the cost. Average wholesale prices of commonly used antiretroviral drugs range from $54 to $4,097 a month.

4.11. HIV AND AIDS: CONNECTION

To develop AIDS, an individual have to have been infected with HIV. But having HIV doesn't necessarily mean you'll develop AIDS. HIV is passed from person to person through bodily fluids such as blood and semen. Once the virus enters your body, it attacks your immune system by destroying CD4 cells, which help keep you from getting sick.

There are three stages of HIV infection:
- acute stage, the first few weeks after infection
- clinical latency, or chronic stage
- AIDS, the last stage

As HIV lowers your CD4 cell count your immune system weakens. A normal adult CD4 count is 800 to 1,000 per cubic millimeter. A count below 200 is considered AIDS.

How quickly HIV progresses through the chronic stage varies significantly from person to person. Without treatment, it can last up to a decade before advancing to AIDS. With treatment, it can last indefinitely. There is no cure for HIV, but it can be controlled. People with HIV often have a near-normal lifespan with early intervention with antiretroviral therapy. There's no cure for AIDS, but individual infections and diseases are often treatable.

REFERENCES

- Sepkowitz KA (June 2001). "AIDS—the first 20 years". N. Engl. J. Med. 344 (23): 1764–72.
- Kallings LO (2008). "The first postmodern pandemic: 25 years of HIV/AIDS". Journal of Internal Medicine. 263 (3): 218–43.
- Vogel, M; Schwarze-Zander, C; Wasmuth, JC; Spengler, U; Sauerbruch, T; Rockstroh, JK (July 2010). "The treatment of patients with HIV". Deutsches Ärzteblatt international. 107 (28–29): 507–15; quiz 516.
- Walker, BD (Aug–Sep 2007). "Elite control of HIV Infection: implications for vaccines and treatment". Topics in HIV medicine : a publication of the International AIDS Society, USA. 15 (4): 134–6.
- Holmes CB, Losina E, Walensky RP, Yazdanpanah Y, Freedberg KA (2003). "Review of human immunodeficiency virus type 1-related opportunistic infections in sub-Saharan Africa". Clin. Infect. Dis. 36 (5): 656–662.
- Chu, C; Selwyn, PA (February 15, 2011). "Complications of HIV infection: a systems-based approach". American family physician. 83 (4): 395–406.
- Sestak K (July 2005). "Chronic diarrhea and AIDS: insights into studies with non-human primates". Curr. HIV Res. 3 (3): 199–205.

- Coovadia H (2004). "Antiretroviral agents—how best to protect infants from HIV and save their mothers from AIDS". N. Engl. J. Med. 351 (3): 289–292.
- Kripke C (1 August 2007). "Antiretroviral prophylaxis for occupational exposure to HIV.". American family physician. 76 (3): 375–6.
- Dosekun O, Fox J (July 2010). "An overview of the relative risks of different sexual behaviours on HIV transmission.". Current opinion in HIV and AIDS. 5 (4): 291–7.
- Beyrer, C; Baral, SD; van Griensven, F; Goodreau, SM; Chariyalertsak, S; Wirtz, AL; Brookmeyer, R (Jul 28, 2012). "Global epidemiology of HIV infection in men who have sex with men". Lancet. 380 (9839): 367–77.

5. EBOLA

Annapurna Sahoo
G. K. Panigrahi

5.1. INTRODUCTION

Ebola virus disease (EVD) or Ebola hemorrhagic fever (EHF) is a viral hemorrhagic fever, often fatal illness in humans and nonhuman primates (monkeys, gorillas, and chimpanzees) caused by ebola viruses. EVD have a case fatality rate of up to 90%. Ebola viruses are found in several African countries. First in 1976, Ebola appeared in two simultaneous outbreaks, in Nzara, Sudan, and in Yambuku, Democratic Republic of Congo. The latter was in a village situated near the Ebola River, from which the disease takes its name. Since then, outbreaks have appeared sporadically in Africa. However, on the basis of evidence and the nature of similar viruses, researchers believe that the virus is animal-borne and that bats are the most likely reservoir. Four of the five subtypes of ebola virus occur in an animal host native to Africa. The risk to human health is likely to be low for healthy adults but is unknown for all other population groups.

5.2. CAUSE AND AGENT

Ebola virus belongs to the Filoviridae family (filo virus), genus Ebola virus. Ebola virus comprises 5 distinct species:

1. *Bundibugyo ebola virus* (BDBV)
2. *Zaire ebola virus* (EBOV)
3. *Sudan ebola virus* (SUDV)
4. *Reston ebola virus* (RESTV)
5. *Taï Forest ebola virus* (TAFV)

Four of the five have caused disease in humans: Ebola virus (*Zaire ebola virus*); Sudan virus (*Sudan ebola virus*); Taï Forest virus (*Taï Forest ebola virus*, formerly *Côte d'Ivoire ebola virus*); and Bundibugyo virus (*Bundibugyo ebolavirus*). Pathogenicity varies among *Ebola viruses*, from EBOV, which is highly lethal in humans, to RESTV, which causes disease in pigs and macaques but asymptomatically infects humans. The RESTV is not as great a threat as the other ebola viruses that are known to be highly pathogenic for humans. The natural reservoir host of Ebola viruses remains unknown.

5.3. DISEASE AGENT CHARACTERISTICS

a) Virion morphology and size: Enveloped, helical, cross-striated nucleocapsid, filamentous or pleomor-phic virions that are flexible with extensive branch-ing, 80 nm in diameter and 970-1200 nm in length.

b) Nucleic acid: Linear, negative-sense, single-stranded RNA, ~18,900 kb in length.

c) Physicochemical properties: Stable at room temperature and can resist desiccation; inactivated at 60°C for 30 minutes; infectivity greatly reduced or destroyed by UV light and gamma irradiation, lipid solvents, β-propiolactone, formaldehyde, sodium hypochlo-rite, and phenolic disinfectants.

5.4. GEOGRAPHICAL DISTRIBUTION

EVD outbreaks occur primarily in remote villages in Central and West Africa, near tropical rainforests. The virus is transmitted to people from wild animals and spreads in the human population through human-to-human transmission. Since 2008, Reston *ebolavirus* has been detected during several outbreaks of a deadly disease in pigs in the People's Republic of China and in Philippines, but no illness or death in humans from this species has been reported to date. On 8 August 2014 the World Health Organisation (WHO) declared the Ebola virus disease (EVD) outbreak in West Africa a Public Health Emergency of International Concern (PHEIC), stressing the need for international attention and collaboration to control the outbreak. At this moment (18 September 2014) a total of 5335 cases with 2622 reported deaths have been notified, in Guinea, Liberia, and Sierra Leone. The imported EVD case in Nigeria that resulted in a relatively small outbreak, and similar imported cases in the USA and Spain which at first appeared to have been well contained, but eventually lead to infection of healthcare workers, show the importance of adequate isolation methods, training of personnel and the adequate use of personal protective equipment (PPE). For the West Africa outbreak, the total number of cases is subject to change due to ongoing reclassification, retrospective investigation and the availability of laboratory results. A second, non-related, EVD outbreak has been reported in the Democratic Republic of Congo with currently a total of 62 confirmed and suspected cases.

5.4.1. RESERVOIR

Despite extensive work, no filovirus vector has been identified. Ape-to-ape transmission may be responsible for the epizootic wave of this disease, although the fruit bat act as a reservoir for this disease. Fruit bats of the Pteropodidae family are well thought-out to be the natural host of the Ebola virus. Although non-human primates have been a source of infection for humans, they are not thought to be the reservoir but rather an accidental host like human beings. Since 1994, Ebola outbreaks from the EBOV and TAFV species have been observed in chimpanzees and gorillas. RESTV has caused severe EVD outbreaks in macaque monkeys (Macaca fascicularis) farmed in Philippines and detected in monkeys imported into the USA in 1989, 1990 and 1996, and in monkeys imported to Italy from Philippines in 1992. A recent study suggests that bats might be a reservoir for *Ebola virus* in Bangladesh. The study found antibodies against Zaire and Reston *ebolaviruses* circulating in 3.5% of the 276 bats scientists screened in Bangladesh. Detection of antibodies to Ebola virus infection in Indonesian orangotans suggests the existence of multiple species of filoviruses or unknown filovirus-related viruses in Indonesia, some of which are serologically similar to African ebolaviruses.

5.5. PATHOGENESIS AND TRANSMISSION

After infection, development of disease is a complex interplay between virus, host and environment. Different case fatality rates (CFR) have been reported between the four human-pathogenic Ebolaviruses. For EBOV the CFR ranges from 50-90% of the EVD cases. For the current outbreak, CFR is estimated to be around 50%, although there is some evidence of improved outcomes with intense symptomatic treatment. *Ebola* is spread through direct contact (through broken skin or unprotected mucous membranes in, for example, the eyes, nose, or mouth) with blood or body fluids (including but not limited to faeces, saliva, sweat, urine, vomit, breast milk, and semen) of a person who is sick with *Ebola*. Infection through intact skin is considered unlikely, although not excluded. The virus has been successfully isolated from skin (biopsy) and body fluids. Objects (like needles and syringes) that have been contaminated with the virus, infected fruit bats or primates (apes and monkeys), and possibly from contact with semen from a man who has recovered from Ebola (for example, by having oral, vaginal, or anal sex). The route of transmission seems to affect the disease outcome. *Ebola* is not spread through the air or by water, or in general, by food. However, in Africa, *Ebola* may be spread as a result of handling bushmeat (wild animals hunted for food) and contact with infected bats. There is no evidence that mosquitos or other insects can transmit *Ebola virus*. Only a few species of mammals (for example, humans, monkeys, and apes) have shown the ability to become infected with and spread *Ebola virus*. In Africa, infection has been documented through the handling of infected chimpanzees, gorillas, fruit bats, monkeys, forest antelope and porcupines found ill or dead or in the rainforest. *Ebola* then spreads in the community through human-to-human transmission, with infection resulting from direct contact (through broken skin or mucous membranes) with the blood, secretions, organs or other bodily fluids of infected people, and indirect contact with environments contaminated with such fluids. Burial ceremonies in which mourners have direct contact with the body of the deceased person can also play a role in the transmission of *Ebola*. Health-care workers have frequently been infected while treating patients with suspected or confirmed EVD. This has occurred through close contact with patients when infection control precautions are not strictly practiced. In the early EBOV outbreak in 1976, CFR after transmission by injection was 100% versus 80% in contact exposure cases. This has been confirmed in a non-human primate model, showing faster disease progression in animals infected via injection versus those that received an aerosol challenge. Post mortem studies of patients and experimentally infected animals showed infection of immune cells (macrophages, monocytes and dendritic cells), epithelial and endothelial cells, fibroblasts, hepatocytes and adrenal gland tissue. Hallmark characteristics of EVD, as in any VHF, are the bleeding manifestations although these are infrequently observed in the current outbreak. Studies addressing the mechanism behind these coagulation abnormalities first showed that haemorrhage was most likely not a direct effect of endothelial cell infection, followed by cytolysis. A more likely explanation seems to be an overexpression of tissue factor in monocytes/macrophages resulting in (over)activation of the extrinsic pathway of coagulation followed by a consumptive coagulopathy and eventually a disseminated intravascular coagulation.

5.6. SIGNS AND SYMPTOMS

EVD is a severe acute viral illness often characterized by the sudden onset of fever, intense weakness, muscle pain, headache and sore throat. Incubation period is usually 2 to 21 days. This is followed by vomiting, diarrhoea, impaired kidney and liver function, and in some cases, both internal and external bleeding. Around Day 5, most patients develop a maculopapular rash that is prominent on the trunk followed by desquamation in survivors. Central nervous system involvement is often manifested by somnolence, delirium, or coma. During the second week, the patient defervesces and improves markedly or dies in shock with multiorgan dysfunction, often accompanied by disseminated intravascular coagulation, anuria, and liver failure. Convalescence may be protracted and accompanied by arthralgia, orchitis, recurrent hepatitis, transverse myelitis, psychosocial disturbances, or uveitis. It also includes low white blood cell and platelet counts and elevated liver enzymes. People are infectious as long as their blood and secretions contain the virus. Men who have recovered from the disease can still transmit the virus through their semen for up to 7 weeks after recovery from illness.

5.7. DIAGNOSIS

Diagnosing *Ebola* in a person infected for only a few days is difficult as the early symptoms, such as fever, are nonspecific to *Ebola* and are seen often in patients with more common diseases, such as malaria and typhoid fever. However, if a person has the early symptoms of *Ebola*, the patient should be secluded and public health professionals notified. Samples from the patient can then be collected and tested to confirm infection. It may take up to three days after symptoms start for the virus to reach detectable levels.

The diagnosis of acute EVD is made by viral genome detection via RT-PCR. The virus is usually detectable 48 hours after infection in both lethal and non-lethal cases. This suggests that a negative test result within the first 48 hours after exposure does not rule out EBOV infection. Due to the quickness of the acute disease, serology does not play a role in diagnosis of acute EVD patients but may be of use in epidemiological and surveillance studies. In common, IgM antibodies can be detected starting from two days after the first symptoms appear and disappear after 30-168 days. IgG response is generally considered to start between day 6 and 18 post onset of illness and remains detectable for years. Antibody profile of the sera from patients with lethal disease as compared with those that survive is markedly distinct. This difference can serve as a predictive marker for the management of the patient since antibody responses strongly differ between lethal and survivor cases and it has been shown that deceased patients show a much lower or even absent antibody response compared with survivors.

5.8. TREATMENT

There is no FDA-approved treatment (e.g., antiviral drug) for *Ebola*. Symptoms and complications are treated as they appear. The following basic intrusions, when used early, can significantly improve the chances of survival:

- Providing intravenous fluids and balancing electrolytes (body salts)
- Maintaining oxygen status and blood pressure
- Treating other infections if they occur

Experimental treatments for *Ebola* are under expansion, but they have not yet been fully tested for safety or effectiveness. Retrieval from *Ebola* depends on good supportive care and the patient's immune response. People who recover from *Ebola* develop antibodies that last for at least 10 years, possibly longer. It isn't known if people who recover are immune for life or if they can become infected with a different species of *Ebola*. Some people who have recovered from *Ebola* have developed long-term complications, such as joint and vision problems. *Ebola* virus has been found in the semen of some men who have recovered from *Ebola*. It is possible that *Ebola* could be spread through sex or other contact with semen. It is not known if *Ebola* can be spread through sex or other contact with vaginal fluids from a woman who has had *Ebola*.

5.9. PREVENTION

While travelling to an area affected by an *Ebola* outbreak, make sure to:

- Practice careful hygiene. For example, wash your hands with soap and water or an alcohol-based hand sanitizer.
- Avoid contact with blood and body fluids.
- Do not handle items that may have come in contact with an infected person's blood or body fluids (such as clothes, bedding, needles, and medical equipment).
- Avoid funeral or burial rituals that require handling the body of someone who has died from *Ebola*.
- Avoid contact with bats and nonhuman primates or blood, fluids, and raw meat prepared from these animals.
- Monitor your health after you return for 21 days and seek medical care immediately if you develop symptoms of *Ebola*.

Healthcare workers who may be exposed to people with Ebola should follow these steps:

- Wear appropriate personal protective equipment (PPE).
- Practice proper infection control and sterilization measures.
- Isolate patients with *Ebola* from other patients.

☐ Avoid direct contact with the bodies of people who have died from *Ebola*.

☐ Notify health officials if you have had direct contact with the blood or body fluids of a person sick with *Ebola*.

Fast and extensive geographic spread of the current Ebola virus out break are motives for increased alertness. Owing to the initial non-specific presentation of EVD, the mixture of fever in combination with high-risk exposure is adequate to proceed with isolation and management protocols in patients who visited endemic areas. Presently, treatment approaches count on solely on the early start of supportive care, where aggressive fluid replacement therapy is proven to drastically improve the survival rates. Precise antiviral EVD treatment approaches are still in the experimental phase. All countries and clinical centres should be conscious of the potential for admission of an EBOV infected person.

REFERENCES

- Isaacson, M; Sureau, P; Courteille, G; Pattyn, SR;. Clinical Aspects of Ebola Virus Disease at the Ngaliema Hospital, Kinshasa, Zaire, 1976.
- WHO. Media Centre. Ebola virus disease. Fact sheet N°103. http://www.who.int/mediacentre/ factsheets/fs103/en/.
- WHO Global Alert and Response. *Ebola* Reston in pigs and humans in the Philippines – update, www. who.int/csr/don/2009_03_31/en/.
- Olival KJ, Islam A, Yu M, Anthony SJ, Epstein JH, Khan SA, Khan SU, Crameri G, Wang LF, Lipkin WI, Luby SP, Daszak P.; Ebola virus antibodies in fruit bats, Bangladesh. Emerg Infect Dis. 2013 Feb;19(2):270-3. doi: 10.3201/eid1902.120524.
- Nidom CA1, Nakayama E, Nidom RV, Alamudi MY, Daulay S, Dharmayanti IN, Dachlan YP, Amin M, Igarashi M, Miyamoto H, Yoshida R, Takada A. Serological evidence of Ebola virus infection in Indonesian orangutans. PLoS One. 2012;7(7): e40740. doi: 10.1371/journal.pone.0040740. Epub 2012 Jul 18.
- M. Goeijenbier, J.J.A. van Kampen, C.B.E.M. Reusken, M.P.G. Koopmans, E.C.M. van Gorp. *Ebola virus* disease: a review on epidemiology, symptoms, treatment and pathogenesis. The Journal of Medicine. November 2 014;72(9):442-448
- Suggested Readings:
- Bermejo M, Rodriguez-Teijeiro JD, Illera G, Barroso A, Vilà C, Walsh PD. Ebola outbreak killed 5000 gorillas. Science 2006;314:1564.
- Chepurnov AA, Bakulina LF, Dadaeva AA, Ustinova EN, Chepurnova TS, Baker JR Jr. Inactivation of *Ebola virus* with a surfactant nanoemulsion. Acta Tropica 2003;87:315-20.
- KsiazekTG,RollinPE, WilliamsAJ,BresslerDS, Martin ML, Swanepoel R, Burt FJ, Leman PA, Khan AS, Rowe AK, Mukunu R, Sanchez A, Peters CJ. Clinical virology

of Ebola hemorrhagic fever (EHF): virus, virus antigen, and IgG and IgM antibody findings among EHF patients in Kikwit, Democratic Republic of the Congo, 1995. J Infect Dis 1999;179 (Suppl 1): S177-87.

- Leroy EM, Kumulungui B, Pourrut X, Rouquet P, Hassanin A, Yaba P, Délicat A, Paweska JT, Gonzalez JP, Swanepoel R. Fruit bats as reservoirs of *Ebola virus*. Nature 2005;438:575-6
- Peters CJ. Marburg and *Ebola virus* hemorrhagic fevers. In: Mandell GL, Bennett JE, Dolin R, editors. Principles and practice of infectious diseases, 6th ed. Philadelphia: Churchill Livingstone; 2005. p. 2057-60.
- Peters CJ, LeDuc JW. An introduction to Ebola: the virus and the disease. J Infect Dis 1999;179 (Suppl1): ix-xvi.
- Rodriguez LL, De Roo A, Guimard Y, Trappier SG, Sanchez A, Bressler D, Williams AJ, Rowe AK, Bertolli J, Khan AS, Ksiazek TG, Peters CJ, Nichol ST. Persistence and genetic stability of *Ebola virus* during the out-break in Kikwit, Democratic Republic of the Congo, 1995. J Infect Dis 1999;179 (Suppl 1):S170-6.
- Sanchez A, Geisbert CW, Feldman H. Filoviridae: Marburg and *Ebola viruses*. In: Knipe DM, Howley PM, editors. Fields virology, 5th ed. Philadelphia: Lippincott, Williams & Wilkins 2007. pp. 1409-48.

6. DENGUE

T. Sarita Achari
T. K. Barik

6.1. INTRODUCTION

Dengue is mosquito borne viral infection which causes flu like illness and occasionally develops into a potentially lethal complication. Common name of dengue is breakbone fever. Dengue fever (DF) and its severe forms-dengue haemorrhagic fever (DHF) and dengue shock syndrome (DSS) have become major international public health concerns. Over the past three decades, there has been a dramatic global increase in the frequency of dengue fever, DHF and DSS and their epidemics, with a parallel increase in disease incidence. Dengue is found in tropical and subtropical regions around the world, predominantly in urban and semi-urban areas. The incidence of dengue has grown dramatically around the world in recent decades. The actual numbers of dengue cases are under reported and many cases are misclassified. One recent estimate indicates 390 million dengue infections per year (95% credible interval 284-528 million), of which 96 million (67–136 million) manifest clinically (with any severity of disease).Another study, of the prevalence of dengue, estimates that 3.9 billion people, in 128 countries, are at risk of infection with dengue viruses (WHO, 2017).The disease is caused by a virus belonging to family Flaviviradae that is spread by *Aedes* mosquitoes. There is no specific treatment for dengue, but appropriate medical care frequently saves the lives of patients with the more serious dengue haemorrhagic fever. The most effective way to prevent dengue virus transmission is to combat the disease-carrying mosquitoes. The transmission of dengue virus depends upon biotic and abiotic factors. Biotic factors include the virus, the vector and the host and the abiotic factors that include temperature, humidity and rainfall.

6.2. ORIGIN AND HISTORY

Dengue fever is an old disease; the first record of a clinically compatible disease being recorded in a Chinese medical encyclopedia in 992. As the global shipping industry expanded in the 18th and 19th centuries, port cities grew and became more urbanized, creating ideal conditions for the principal mosquito vector, Aedes aegypti. Both the mosquitoes and the viruses were thus spread to new geographic areas causing major epidemics. Because dispersal was by sailing ship, however, there were long intervals (10-40 years) between epidemics. In the aftermath of World War II, rapid urbanization in Southeast Asia led to increased transmission and hyperendemicity. The first major epidemics of the severe and fatal form of disease, dengue haemorrhagic fever, occurred in Southeast Asia as a direct result of this changing ecology. In the last 25 years of the 20th century, a dramatic global geographic expansion of epidemic DF/DHF occurred, facilitated by unplanned urbanization in tropical developing countries, modern transportation, lack of effective mosquito control and globalization.

6.3. DISTRIBUTION

Dengue is found in tropical and sub-tropical climates worldwide, mostly in urban and semi-urban areas. Severe dengue is a leading cause of serious illness and death among children in some Asian and Latin American countries (WHO, 2017).

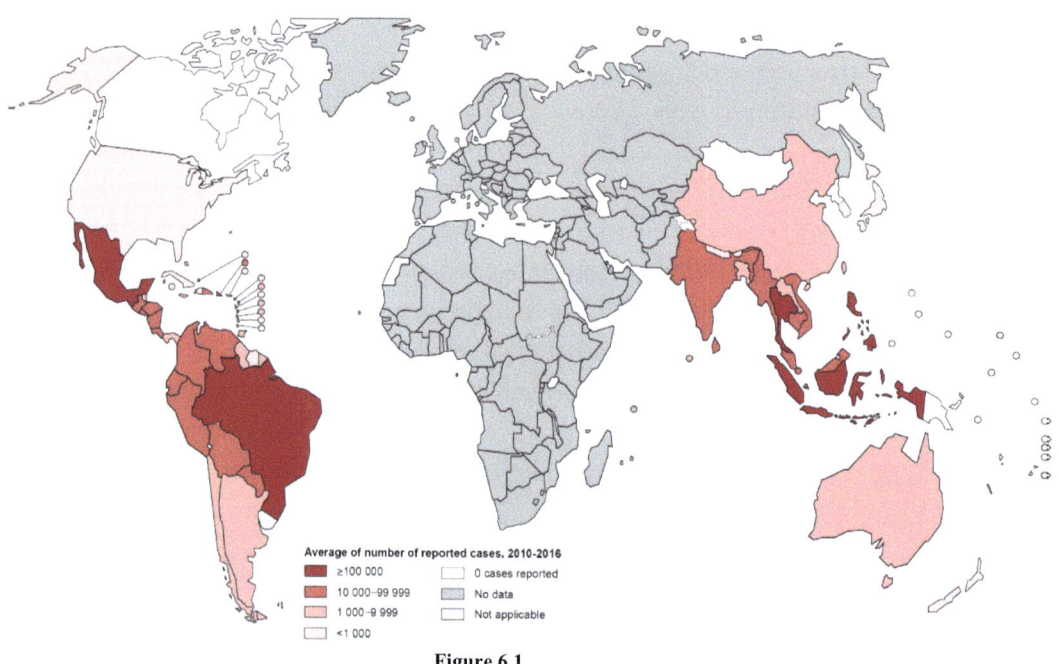

Figure 6.1

Surce: distribution of dengue, worldwide-2016

6.4. STRUCTURE AND CHARACTERISTICS OF DENGUE VIRUS

The dengue viruses are members of the genus Flavivirus and family Flaviviridae. These small (50nm) viruses contain single-strand RNA as genome. The virion consists of a nucleocapsid with cubic symmetry enclosed in a lipoprotein envelope. The dengue virus genome is 11644 nucleotides in length, and is composed of three structural protein genes encoding the nucleocapsid or core protein (C), a membrane-associated protein (M), an envelope protein (E), and seven non-structural protein (NS) genes. Among non-structural proteins, envelope glycoprotein, NS1, is of diagnostic and pathological importance. It is 45 kDa in size and associated with viral haemagglutination and neutralization activity. The dengue viruses form a distinct complex within the genus Flavivirus based on antigenic and biological characteristics. There are four virus serotypes, which are designated as DENV-1, DENV-2, DENV-3 and

DENV-4. Infection with any one serotype confers lifelong immunity to that virus serotype. Although all four serotypes are antigenically similar, they are different enough to elicit cross-protection for only a few months after infection by any one of them. Secondary infection with another serotype or multiple infections with different serotypes leads to severe form of dengue (DHF/DSS).

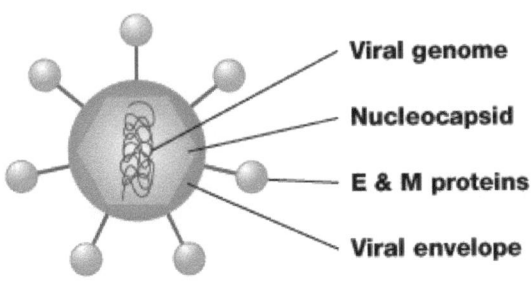

Figure 6.2

Dengue virus showing structural components
Source:http://bestpractice.bmj.com/bestactice/monograph/1197/basics/pathophysiology.html.

6.5. REPLICATION OF DENGUE VIRUS

The life cycle of the dengue involves endocytosis via a cell surface receptor. The virus uncoats intracellularly via a specific process. In the infectious form of the virus, the envelope protein lays flat on the surface of the virus, forming a smooth coat with icosahedral symmetry. However, when the virus is carried into the cell and into lysozomes, the acidic environment causes the protein to snap into a different shape, assembling into trimeric spike. Several hydrophobic amino acids at the tip of this spike insert into the lysozomal membrane and cause the virus membrane to fuse with lysozome. This releases the RNA into the cell and infection starts. The DENV RNA genome is in the infected cell translated by the host ribosomes. The resulting polyprotein is subsequently cleaved by cellular and viral proteases at specific recognition sites. The viral non-structural proteins use a negative-sense intermediate to replicate the positive-sense RNA genome, which then associates with capsid protein and is packaged into individual virions. Replication of all positive-stranded RNA viruses occurs in close association with virus-induced intracellular membrane structures. DENV also induces such extensive rearrangements of intracellular membranes, called replication complex (RC). These RCs seem to contain viral proteins, viral RNA and host cell factors. The subsequently formed immature virions are assembled by budding of newly formed nucleocapsids into the lumen of the endoplasmic reticulum (ER), thereby acquiring a lipid bilayer envelope with the structural proteins prM (the precursor of membrane) and E. The virions mature during transport through the acidic trans-Golgi network, where the prM proteins stabilize the E proteins to prevent conformational changes. Before release of the virions from the host cell, the maturation process is completed when prM is cleaved into a soluble pr (precursor) peptide and virion-associated M by the cellular protease furin. Outside the cell, the virus particles encounter a neutral pH, which promotes dissociation of the pr peptides from the virus particles and generates mature, infectious virions. At this point the cycle repeats itself.

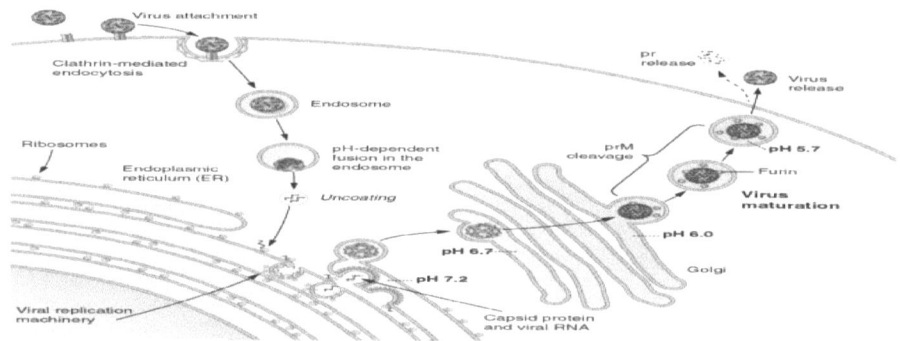

Figure 6.3

Dengue virus life cycle.
Source http://www.denguevirusnet.com/dengue-virus.html

6.6. VECTORS OF DENGUE

The primary vector of Dengue is *Aedes aegypti mosquito*, preferring to live in and around homes in tropical and subtropical regions. This mosquito feeds preferentially on human blood under field conditions, with the geographic range spanning all continents except Antarctica. A secondary dengue vector, *Aedes albopictus*, living out doors, but still feeds almost exclusively on humans. The strong preferences for human blood exhibited by these mosquitoes increase the potential for disease transmission among humans. These mosquitoes are day biter. The immature stages are found in water-filled habitats, mostly in artificial containers closely associated with human dwellings and often indoors.

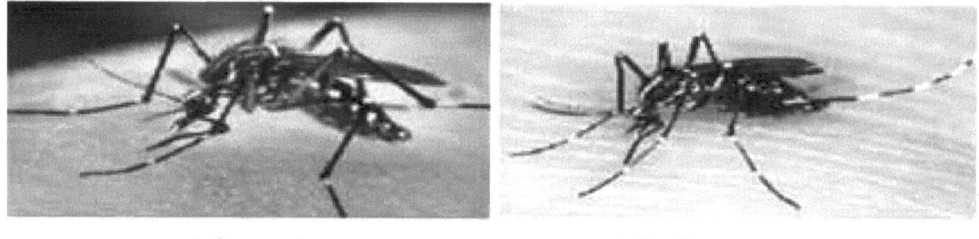

Aedes aegypti   *Aedes albopictus*

Figure 6.4

Source:http://www.who.int/immunization/research/development/dengue_vaccines/en/ .

6.7. HOST FACTOR

Dengue virus infects humans and several species of lower primates. In India, man is the only natural reservoir of infection. All ages and both sexes are susceptible to dengue fever. Secondary dengue infection is a risk factor for DHF including passively acquired antibodies in infants. Travel to dengue endemic area is an important risk factor, if the patient develops fever more than 2 weeks after travel, dengue is unlikely. Migration of patient during viremia to a non-endemic area may introduce it into the area.

6.8. TRANSMISSION

Humans are the main amplifying host of the dengue virus. Dengue virus circulating in the blood of viraemic humans is ingested by female *Aedes* mosquitoes during feeding. The virus then infects the mosquito mid-gut and subsequently spreads systemically over a period of 8-12 days. After this extrinsic incubation period, the virus can be transmitted to other humans during subsequent probing or feeding. The extrinsic incubation period is influenced in part by environmental conditions, especially ambient temperature. Thereafter the mosquito remains infective for the rest of its life. *Ae. aegypti* is one of the most efficient vectors for arboviruses because it is highly anthropophilic, frequently bites several times before completing oogenesis, and thrives in close proximity to humans. Several factors can influence the dynamics of virus transmission including environmental and climate factors, host-pathogen interactions and population immunological factors.

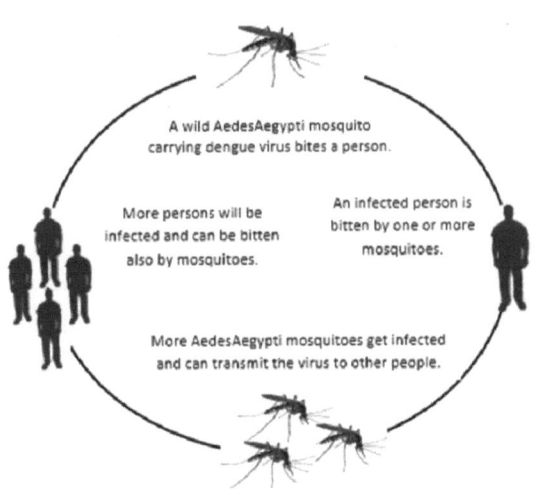

Figure 6.5

Transmission cycle of dengue
Source:http://www.who.int/immunization/research/development/dengue_vaccines/en/ .

6.9. CLINICAL PRESENTATION

Dengue virus infection may be asymptomatic or may cause undifferentiated febrile illness (viral syndrome), dengue fever (DF), or dengue haemorrhagic fever (DHF) including dengue shock syndrome (DSS). Infection with one dengue serotype gives lifelong immunity to that particular serotype, but there is no cross protection for the other serotypes. The clinical presentation depends on age, immune status of the host, and the virus strain.

(I) Undifferentiated fever

Infants, children and some adults who have been infected with dengue virus for the first time (i.e. primary dengue infection) will develop a simple fever indistinguishable from other viral infections. Maculopapular rashes may accompany the fever or may appear during defervescence.

(II) Dengue fever

Dengue fever is most common in older children and adults. It is generally an acute biphasic fever with headache, myalgias, arthralgias, rashes and leucopenia. Although DF is commonly benign, it may be an incapacitating disease with severe muscle and joint pain (break-bone fever), particularly in adults, and occasionally with unusual haemorrhage. In dengue endemic areas, DF seldom occurs among indigenous people.

(III) Dengue haemorrhagic fever

Dengue haemorrhagic fever is most common in children less than 15 years of age, but it also occurs in adults. DHF is characterized by the acute onset of fever and associated non-specific constitutional signs and symptoms. There is a haemorrhagic diathesis and a tendency to develop fatal shock (dengue shock syndrome). Abnormal haemostasis and plasma leakage are the main pathophysiological changes, with thrombocytopenia and haemoconcentration presenting as constant findings. Although DHF occurs most commonly in children who have experienced secondary dengue infection, it has also been documented in primary infections.

(IV) Dengue shock syndrome (DSS)

Dengue shock syndrome (DSS) is a form of hypovolaemic shock and results from continued vascular permeability and plasma leakage. This usually takes place around defervescence, i.e. on days 4−5 of illness (range of days 3−8), and is often preceded by warning signs. From this point onwards, patients who do not receive prompt intravenous fluid therapy progress rapidly to a state of shock. Dengue shock presents as a physiologic continuum, progressing from asymptomatic capillary leakage to compensated shock to hypotensive shock and ultimately to cardiac arrest.

6.10. SYMPTOMS

After an average incubation period of 7 days (range 3-14 days), various non-specific, undifferentiated prodomes, such as headache, backache and general malaise may develop. Typically, the onset of DF in adults is sudden, with a sharp rise in temperature occasionally accompanied by chillis, and is invariably associated with severe headache and flushed face.

Within 24 hours there may be retroorbital pain, particularly on eye movement or eye pressure, photophobia, backache and pain in the muscles and joints/bones of the extremities. The other common symptoms include anorexia and altered taste sensation, constipation, colicky pain and abdominal tenderness, dragging pains in the inguinal region, sore throat, and general depression. These symptoms vary in severity and usually persist for several days.

- **Fever:** The body temperature is usually between 39°C and 40°C, and the fever may be biphasic, lasting for 5-7 days.
- **Rash:** Diffuse flushing or fleeting pinpoint eruptions may be observed on the face, neck and chest during the first half of the febrile period, and a conspicuous rash that may be maculopapular or scarlatiniform appears on approximately the third or fourth day. Towards the end of the febrile period or immediately after defervescence, the generalized rash fades and localized clusters of petechiae may appear over the dorsum of the feet, on the legs, and on the hands and arms. This confluent petechial rash is characterized by scattered, pale, round areas of normal skin. Occasionally the rash is accompanied by itching.
- **Skin Haemorrhage**: Bleeding into the skin can occur from broken blood vessels that form tiny red dots (called petechiae). Blood also can collect under the tissue in larger flat areas (called purpura), or in a very large bruised area (called an ecchymosis).
- **Course**: The relative duration and severity of DF varies between individuals in a given epidemic, as well as from one epidemic to another. Convalescence may be short and uneventful, but may also often be prolonged. In adults it sometimes lasts for several weeks and may be accompanied by pronounced asthenia and depression. Bradycardia is common during convalescene. Haemorrhagic complications, such as epistaxis, gingival bleeding, gastrointestinal bleeding, haematuria and hypermenorrhoea, may accompany epidemics of DF. Severe bleeding has occasionally caused deaths in some epidemics.

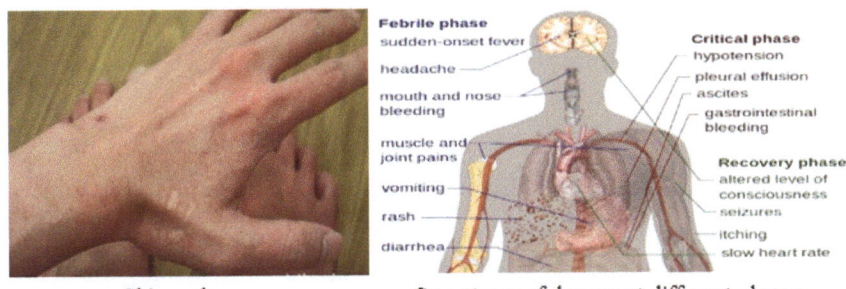

Skin rashes   Symptoms of dengue at different phases

Figure 6.6

Source: http://www.who.int/immunization/research/development/dengue_vaccines/en/ .

6.11. PATHOPHYSIOLOGY

Pathogenesis is linked to the host immune response, which is triggered by infection with the dengue virus. Primary infection is usually benign in nature; however, secondary infection with a different serotype or multiple infections with different serotypes may cause severe infection that can be classified as either dengue haemorrhagic fever (DHF) or dengue shock syndrome (DSS), depending on the clinical signs. The incubation period is 3 to 14 days (average 7 days). Non-neutralising, cross-reactive antibodies elicited by a previous primary infection are involved in the phenomenon of antibody-dependent enhancement (ADE), which causes a heavy viral burden. Cells of the monocyte-macrophage lineage are the major sites of viral replication, but the virus can infect other tissues in the body such as the liver, brain, pancreas, and heart. Antigen-presenting dendritic cells, the humoral immune response, and the cell-mediated immune response are involved in the pathogenesis. Proliferation of memory T cells and the production of pro-inflammatory cytokines leads to vascular endothelial cell dysfunction, which results in plasma leakage. There is a higher concentration of cytokines such as interferon-gamma, tumour necrosis factor (TNF)-alpha, and interleukin-10, as well as reduced levels of nitric oxide and some complement factors. NS1 is a modulator of the complement pathway and plays a role in low levels of complement factors. After infection, specific cross-reactive antibodies, as well as CD4+ and CD8+ T cells, remain in the body for years. Vascular leak is the hallmark of DHF and DSS, and causes an increase in haematocrit, hypoalbuminaemia, and the development of pleural effusions or ascites. Preliminary data suggest that transient dysfunction of the endothelial glycocalyx layer leads to vascular leak. There is also a tendency towards haemorrhage associated with severe thrombocytopenia and coagulation disorders. In severe infection, loss of intravascular fluid leads to tissue hypoperfusion, resulting in lactic acidosis, hypoglycaemia, hypocalcaemia, and, finally, multiple organ failure. Multiple organ dysfunction including myocarditis, encephalopathy, and liver cell necrosis can also result from direct viral damage to, and subsequent inflammation in, tissues. Infants can develop severe dengue infection during a primary infection (which is usually benign in nature) due to transplacental transfer of maternal antibodies from an immune mother, which subsequently amplifies the infant's immune response to the primary infection.

6.12. DIAGNOSIS

6.12.1. Clinical laboratory findings

In dengue endemic areas, positive tourniquet test and leukopenia (WBC ≤5000 cells/mm3) help in making early diagnosis of dengue infection with a positive predictive value of 70%–80%. The laboratory findings during an acute DF episode of illness are as follows.

Total WBC is usually normal at the onset of fever; then leucopenia develops with decreasing neutrophils and lasts throughout the febrile period. Platelet counts are usually normal, as are other components of the blood clotting mechanism. Mild thrombocytopenia (100 000 to 150 000 cells/mm3) is common and about half of all DF patients have platelet count below 100 000 cells/mm3. Mild haematocrit rise (≈10%) may be found as a consequence of dehydration associated with high fever, vomiting, anorexia and poor oral intake. Serum biochemistry is usually normal but liver enzymes and aspartate amino transferase (AST) levels may be elevated. It should be noted that the use of medications such as analgesics, antipyretics, anti-emetics and antibiotics can interfere with liver function and blood clotting.

6.12.2. LABORATORY DIAGNOSIS

Early symptoms of dengue fever mimic other diseases often prevalent in areas where it is endemic, such as chikungunya, malaria and leptospirosis. Hence for proper management rapid differential diagnosis is very crucial. Laboratory diagnosis can be carried out by following procedures:

- Isolation of Dengue virus from serum, plasma, leucocytes or autopsy samples.
- Demonstration of a fourfold or greater rise in reciprocal IgG antibody titres to one or more dengue virus antigen in paired sera samples.
- Demonstration of dengue virus antigen in autopsy tissue by immunohistochemistry or immunofluorescence.
- Detection of viral genomic sequences in autopsy tissue, serum or CSF sample by PCR (Polymerase Chain Reaction).

1.1.1.1 Isolation of *Dengue Virus*

Isolation of most strains of dengue virus from clinical specimens can be accomplished in a majority of cases provided the sample is taken in the first few days of illness and processed without delay. Specimens that may be suitable for virus isolation include acute phase serum, plasma or washed buffy coat from the patient, autopsy tissues from fatal cases, especially liver, spleen, lymph nodes and thymus, and mosquitoes collected in nature. This method is suitable for research or for other academic purpose not for patient care.

1.1.1.2 Serological Tests

Following tests are available for the diagnosis of dengue infection, Haemagglutination-Inhibition (HI), Complement Fixation (CF), Neutralization test (NT), IgM capture enzyme-linked immunosorbent assay (MAC-ELISA), and Indirect IgG ELISA, RDT.

1.1.1.3 Collection of Specimen

Laboratory diagnosis of dengue depends on proper collection, processing, storage and shipment of the specimens. While collecting blood for serological studies from suspected DF/DHF cases all universal precautions should be taken.

6.13. TREATMENT

The mainstay of treatment is timely supportive therapy to tackle shock due to hem concentration and bleeding. Close monitoring of vital signs in critical period (between day 2 to day 7 of fever) is critical. Increased oral fluid intake is recommended to prevent dehydration. Supplementation with intravenous fluids may be necessary to prevent dehydration and significant concentration of the blood if the patient is unable to maintain oral intake. A platelet transfusion is indicated in rare cases if the platelet level drops significantly (below 20,000) or if there is significant bleeding. The presence of melena may indicate internal gastrointestinal bleeding requiring platelet and/or red blood cell transfusion. People who suffer from dengue fever have no risk of death but some of them develop Dengue Hemorrhagic Fever (DHF) or Dengue Shock Syndrome (DSS). In some of these cases death can occur. If a clinical diagnosis is made early, a health care provider can effectively treat DHF using fluid replacement therapy. Adequately management of DHF generally requires hospitalization.

6.13.1. Medication

Aspirin, Brufen and non-steroidal anti-inflammatory drugs should be avoided as these drugs may worsen the bleeding tendency associated with some of these infections. Patients may receive paracetamol preparations to deal with these symptoms if dengue is suspected. Any medicines that decrease platelets should be avoided.

6.13.2. Immunization

In late 2015 and early 2016, the first dengue vaccine, Dengvaxia (CYD-TDV) by Sanofi Pasteur, was registered in several countries for use in individuals 9-45 years of age living in endemic areas. Other tetravalent live-attenuated vaccines are under development in phase III clinical trials, and other vaccine candidates (based on subunit, DNA and purified inactivated virus platforms) are at earlier stages of clinical development (Immunization, Vaccines and Biologicals-WHO).

6.14. PREVENTION

Prevention depends on control of and protection from the bites of the mosquito that transmits it. preventing mosquitoes from accessing egg-laying habitats by environmental management and modification; disposing of solid waste properly and removing artificial man-made habitats; covering, emptying and cleaning of domestic water storage containers on a weekly basis; applying appropriate insecticides to water storage outdoor containers; using of personal household protection such as window screens, long-sleeved clothes, insecticide treated materials, coils and vaporizers ; improving community participation and mobilization for sustained vector control; applying insecticides as space spraying during outbreaks as one of the emergency vector-control measures; active monitoring and surveillance of vectors should be carried out to determine effectiveness of control interventions.

REFERENCES

- WHO ,Dengue and severe dengue Fact sheet-2017.
- Duijl-Richter, M. (2016). Dengue and Chikungunya virus: Cell entry mechanisms and the impact of antibodies on infectivity [Groningen]: University of Groningen.
- http://www.who.int/immunization/research/development/dengue_vaccines/en/ .
- http://bestpractice.bmj.com/bestactice/monograph/1197/basics/pathophysiology.html.

7. CHOLERA

Sasmita Panda

7.1. INTRODUCTION

Cholera is an acute epidemic infectious disease. It is characterized by watery diarrhoea, extreme loss of fluid and electrolytes and severe dehydration. It can be fatal. It is caused by the bacterium *Vibrio cholera (V. Cholera)*. Despite being easy to treat, cholera is estimated to affect between 3 and 5 million people each year, and it causes over 100,000 deaths worldwide. Due to severe dehydration, fatality rates are high when untreated, especially among children and infants. Death can occur in otherwise healthy adults within hours. Those who recover usually have long-term immunity against re-infection. Cholera was prevalent in the United States in the 1800s, but now it is rare, because there are well-developed sanitary systems and living conditions. When traveling to Asia, Africa and some parts of Latin America, however, people need to protect themselves against cholera by having the appropriate vaccinations beforehand, drinking only water that is boiled or from a sealed bottle, and following good hand washing practices.

The first cholera pandemic occurred in the Bengal region of India, near Calcutta starting in 1817 through 1824. The disease dispersed from India to Southeast Asia, the Middle East, Europe, and Eastern Africa through trade routes. The second pandemic lasted from 1827 to 1835 and particularly affected North American and Europe due to the result of advancements in transportation and global trade, and increased human migration, including soldiers. The third pandemic erupted in 1839, persisted until 1856, extended to North Africa, and reached South America, for the first time specifically affecting Brazil. The fourth pandemic lasted from 1863 to 1875 spread from India to Nepals and Spain. The fifth pandemic was from 1881-1896 and started in India and spread to Europe, Asia, and South America. The sixth pandemic started 1899–1923. These epidemics were less fatal due to a greater understanding of the cholera bacteria. Egypt, the Arabian peninsula, Persia, India, and the Philippines were hit hardest during these epidemics, while other areas, like Germany in 1892 and Naples from 1910–1911, also experienced severe outbreaks. The final pandemic originated in 1961 in Indonesia and is marked by the emergence of a new strain, nicknamed El Tor, which still persists today in developing countries.

Cholera affects an estimated 3–5 million people worldwide, and causes 58,000–130,000 deaths a year as of 2010. This occurs mainly in the developing world. In the early 1980s, death rates are believed to have been greater than 3 million a year. It is difficult to calculate exact numbers of cases, as many go unreported due to concerns that an outbreak may have a negative impact on the tourism of a country. Cholera remains both epidemic and endemic in many areas of the world.

Although much is known about the mechanisms behind the spread of cholera, this has not led to a full understanding of what makes cholera outbreaks happen in some places and not others.

Lack of treatment of human faeces and lack of treatment of drinking water greatly facilitate its spread, but bodies of water can serve as a reservoir, and seafood shipped long distances can spread the disease. Cholera was not known in the Americas for most of the 20th century, but it reappeared towards the end of that century.

7.2. SYMPTOMS

Only around 1 in 20 cholera infections are severe, and a high percentage of infected people show no symptoms. If symptoms appear, they will do so between 12 hours and 5 days after exposure. They range from mild or asymptomatic to severe. They typically include:

Large volumes of explosive watery diarrhoea, sometimes called "rice water stools" because it can look like water that has been used to wash rice

- Vomiting
- Leg cramps
- A person with cholera can quickly lose fluids, up to 20 liters a day, so severe dehydration and shock can occur. Signs of dehydration include:
- Loose skin
- Sunken eyes
- Dry mouth
- Decreased secretion, for example, less sweating
- Fast heart beat
- Low blood pressure
- Dizziness or lightheadedness
- Rapid weight loss
- Shock can lead to collapse of the circulatory system. It is a life-threatening condition and a medical emergency.

7.3. CAUSE

Cholera bacteria Vibrio cholera enters the body through the mouth, often in food or water that has been contaminated with human waste, due to poor sanitation and hygiene. They can also enter by eating seafood that is raw or not completely cooked, in particular shellfish native to estuary environments, such as oysters or crabs. Poorly cleaned vegetables irrigated by contaminated water sources are another common source of infection. In situations where sanitation is severely challenged, such as in refugee camps or communities with highly limited water resources, a single affected victim can contaminate all the water for an entire population.

7.4. MECHANISM OF TRANSMISSION

Cholera has been found in two animal populations: shell fish and plankton. Transmission is usually through the fecal-oral route of contaminated food or water caused by poor sanitation. Most cholera cases in developed countries are a result of transmission by food, while in the developing world it is more often water. Food transmission can occur when people harvest seafood such as oysters in waters infected with sewage, as *Vibrio cholerae* accumulates in planktonic crustaceans and the oysters eat the zooplankton.

People infected with cholera often have diarrhea, and disease transmission may occur if this highly liquid stool, colloquially referred to as "rice-water", contaminates water used by others. The source of the contamination is typically other cholera sufferers when their untreated diarrheal discharge is allowed to get into waterways, ground water or drinking water supplies. Drinking any infected water and eating any foods washed in the water, as well as shell fish living in the affected water way, can cause a person to contract an infection. Cholera is rarely spread directly from person to person. Both toxic and non-toxic strains exist. Non-toxic strains can acquire toxicity through a temperate bacteriophage.

When consumed, most bacteria do not survive the acidic conditions of the human stomach. The few surviving bacteria conserve their energy and stored nutrients during the passage through the stomach by shutting down much protein production. When the surviving bacteria exit the stomach and reach the small intestine, they must propel themselves through the thick mucus that lines the small intestine to reach the intestinal walls where they can attach and thrive.

Once the cholera bacteria reach the intestinal wall they no longer need the flagella to move. The bacteria stop producing the protein flagellin to conserve energy and nutrients by changing the mix of proteins which they express in response to the changed chemical surroundings. On reaching the intestinal wall, *V. cholerae* start producing the toxic proteins that give the infected person a watery diarrhoea. This carries the multiplying new generations of *V. cholerae* bacteria out into the drinking water of the next host if proper sanitation measures are not in place.

7.5. DIAGNOSIS

A doctor may suspect cholera if a patient has severe watery diarrhoea, vomiting and rapid dehydration, especially if they have recently travelled to a place that has a recent history of cholera, or poor sanitation, or if they have recently consumed shellfish. A stool sample will be sent to a laboratory for testing, but if cholera is suspected, the patient must begin treatment even before the results come back.

7.6. TREATMENT

It is normally dehydration that leads to death from cholera, so the most important treatment is to give oral hydration solution (ORS), also known as oral rehydration therapy (ORT). The treatment consists of large volumes of water mixed with a blend of sugar and salts. Prepackaged mixtures are commercially available, but widespread distribution in developing countries is limited by cost, so home made ORS recipes are often used, with common household ingredients. Severe cases of cholera require intravenous fluid replacement. An adult weighing 70 kilograms will need at least 7 liters of intravenous fluids. Antibiotics can shorten the duration of the illness, but the WHO does not recommend mass use of antibiotics for cholera, because of the growing risk of bacterial resistance. Anti-diarrhoeal medicines are not used because they prevent the bacteria from being flushed out of the body. With proper care and treatment, the fatality rate should be around 1 percent.

7.7. PREVENTION

Cholera is often spread through food and because of poor hygiene. Some simple measures can reduce the risk of contracting cholera. When traveling in areas where the disease is endemic, it is important to:

- Eat only fruit you have peeled.
- Avoid salads, raw fish, and uncooked vegetables.
- Ensure that food is thoroughly cooked.
- Make sure water is bottled or boiled and safe to consume.
- Avoid street food, as this can carry cholera and other diseases.

Travellers should learn about cholera before visiting a country where it is prevalent. Individuals should seek medical attention immediately if they experience symptoms such as leg cramps, vomiting, and diarrhea while in a community where the disease exists. Surveillance and prompt reporting allow for containing cholera epidemics rapidly. Cholera exists as a seasonal disease in many endemic countries, occurring annually mostly during rainy seasons. Surveillance systems can provide early alerts to outbreaks, therefore leading to co-ordinated response and assist in preparation of preparedness plans. Efficient surveillance systems can also improve the risk assessment for potential cholera outbreaks. Understanding the seasonality and location of outbreaks provides guidance for improving cholera control activities for the most vulnerable. For prevention to be effective, it is important that cases be reported to national health authorities.

7.8. VACCINE

There are currently three cholera vaccines recommended by the World Health Organization (WHO). These are **Dukoral, Shanchol**, and **Euvichol**. All three require two doses to give full protection. Dukoral needs to be taken with clean water, and it provides roughly 65 percent protection for 2 years. Shanchol and Euvichol do not need to be taken with water, and they provide 65 percent protection for 5 years. All the vaccines offer higher protection nearer to the time they are given.

REFERENCES

- Azman AS, Rudolph KE, Cummings DA, Lessler J (November 2012). "The incubation period of cholera: A systematic review". Journal of Infection. 66 (5): 432–438.
- Harris, JB; LaRocque, RC; Qadri, F; Ryan, ET; Calderwood, SB (30 June 2012). "Cholera.". Lancet. 379 (9835): 2466–76.
- King AA, Ionides EL, Pascual M, Bouma MJ (August 2008). "Inapparent infections and cholera dynamics". Nature. 454 (7206): 877–80.
- Lozano R, Naghavi M, Foreman K, Lim S, Shibuya K, Aboyans V, Abraham J, Adair T, Aggarwal R, Ahn SY, et al. (December 15, 2012). "Global and regional mortality from 235 causes of death for 20 age groups in 1990 and 2010: a systematic analysis for the Global Burden of Disease Study 2010". Lancet. 380 (9859): 2095–128.
- O'Neal CJ, Jobling MG, Holmes RK, Hol WG (2005). "Structural basis for the activation of cholera toxin by human ARF6-GTP". Science. 309 (5737): 1093–6.
- Reidl J, Klose KE (June 2002). "*Vibrio cholerae* and cholera: out of the water and into the host". FEMS Microbiol. Rev. 26 (2): 125–39.
- Renbourn, E. T. (2012). "The History of the Flannel Binder and Cholera Belt". Medical History. 1 (3): 211–25.
- Sack DA, Sack RB, Chaignat CL (August 2006). "Getting serious about cholera". N. Engl. J. Med. 355 (7): 649–51.
- Sack DA, Sack RB, Nair GB, Siddique AK (January 2004). "Cholera". Lancet. 363 (9404): 223–33.
- Sehdev PS (November 2002). "The origin of quarantine". Clin. Infect. Dis. 35 (9): 1071–1072.
- Telmesani AM (May 2010). "Oral rehydration salts, zinc supplement and *rota virus* vaccine in the management of childhood acute diarrhea". Journal of family and community medicine. 17 (2): 79–82.

8. LEPROSY

T.K. Barik
Deepika Panda
Sasmita Panda

8.1. INTRODUCTION

The bacterium *Mycobacterium leprae* causing Leprosy is a chronic, progressive bacterial infection. It is also known as Hansen's disease after the name of the scientist Dr Gerhard Armauer Hansen, the Norwegian scientist who first discovered Mycobacterium leprae in 1873. It is called kusht in Hindi (India), kusta in Indonesian and rate in Tetum (Timor-Leste). It primarily affects the peripheral nerves, the lining of the nose, eyes and the upper respiratory tract. Leprosy produces skin sores, nerve damage, and muscle weakness. It is curable and treatment in early stage can prevent disability. It is one of the oldest diseases in recorded history. According to the World Health Organization (WHO), the first known written reference to leprosy is from 600 B.C. It is common in many countries, especially those with tropical or subtropical climates.

8.2. HISTORY

Leprosy is an age-old disease, described in the literature of ancient civilizations. The first breakthrough occurred in the 1940s with the development of the drug dapsone. The duration of the treatment was many years, often a lifetime. In the 1960s, M. leprae started to develop resistance to dapsone, the world's only known anti-leprosy drug at that time. The search for further effective antileprosy drugs led to the use of clofazimine and rifampicin in the 1960s and 1970s. Later, Indian scientist Shantaram Yawalkar and his colleagues formulated a combined therapy using rifampicin and dapsone, intended to mitigate bacterial resistance. MDT combining all three drugs was first recommended by the WHO in 1981.These three antileprosy drugs are still used in the standard MDT regimens. In 1981, a WHO Study Group recommended MDT free of cost to all leprosy patients in the world.

8.3. GEOGRAPHICAL DISTRIBUTION

The distribution of new leprosy cases by country among 136 countries that reported to WHO in 2015. India reported 127 326 new cases, accounting for 60% of the global new leprosy cases; Brazil, reported 13% of the global new cases; and Indonesia reported 8% of the global case load. Collectively, these countries reported 14% of all new cases globally. The remaining 5% were reported by 92 countries. Thirty countries reported zero new cases. Ninety-two countries did not report cases of leprosy.

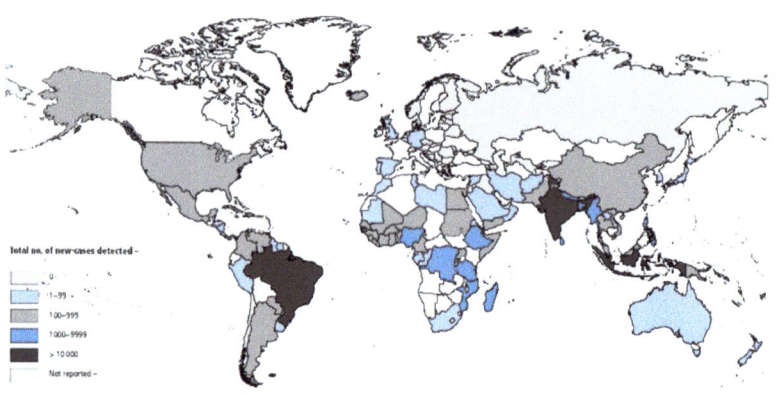

Figure 8.1

Source: Centre for Disease Control

8.4. CAUSATIVE AGENT

M. leprae and *M. lepromatosis* are the causative agents of leprosy. M. lepromatosis is a relatively newly identified mycobacterium isolated from a fatal case of diffuse lepromatous leprosy in 2008. *M. leprae* is an obligate intracellular bacillus, acid-fast, aerobic, gram-positive, rod-shaped bacterium, and is surrounded by the waxy cell membrane coating characteristic of the *Mycobacterium genus*. These are slightly curved, measure from 1 to 8m in length and 0.3m in diameter; like other mycobacteria, they replicate by binary fission.

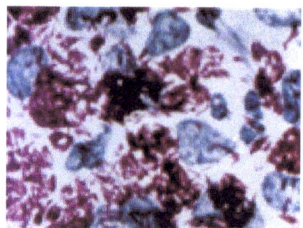

Figure 8.2

Picture of Mycobacterium leprae
Source: .(http://www.who.int/lep/mleprae.jpg)

8.5. TYPES OF LEPROSY

There are three systems for classifying leprosy. The first system recognizes two types of leprosy: tuberculoid and lepromatous. A person's immune response to the disease determines the type of leprosy.

In tuberculoid leprosy, the immune response is good and the disease only exhibits a few lesions (sores on the skin). The disease is mild and only mildly contagious.

In *lepromatous leprosy*, the immune response is poor and affects the skin, nerves, and other organs. There are widespread lesions and nodules (large lumps and bumps). This disease is more contagious.

Basing on the type and number of affected skin areas WHO categorizes the disease into 2 types. The first category is paucibacillary, in which five or fewer lesions with no bacteria are detected in the skin sample. The second category is multibacillary, in which there are more than five lesions, bacteria is detected in the skin smear, or both.

Basing on severity of symptoms Clinical studies use the Ridley-Jopling system, it has six classifications. Viz.,

1) Intermediate leprosy: a few flat lesions that sometimes heal by themselves and can progress to a more severe type.
2) Tuberculoid leprosy: a few flat lesions, some large and numb; some nerve involvement; can heal on its own, persist, or may progress to a more severe form.
3) Borderline tuberculoid leprosy: lesions similar to tuberculoid but smaller and more numerous; less nerve enlargement; may persist, revert to tuberculoid, or advance to another form.
4) Mid-borderline leprosy: reddish plaques, moderate numbness, swollen lymph glands; may regress, persist, or progress to other forms.
5) Borderline lepromatous leprosy: many lesions including flat lesions, raised bumps, plaques, and nodules, sometimes numb; may persist, regress, or progress.
6) Lepromatous leprosy: many lesions with bacteria; hair loss; nerve involvement; limb weakness; disfigurement; doesn't regress.

8.6. TRANSMISSION

The respiratory tract plays a significant role in transmission has recently been studied. The bacterial load is high in patients with lepromatous leprosy, who have been reported to harbor as many as 7000 million bacilli in a gram of tissue, whereas in other forms of the disease, the load is known to be much lower. *M leprae* has been found in high numbers (100 million viable bacilli per day) in nasal mucosa. This usually occurs when the infected person sneezes or coughs. The skin also has been suggested to be a possible route of transmission. The disease isn't highly contagious. Close, repeated contact with an untreated person can lead to contracting leprosy. The bacteria responsible for leprosy multiply very slowly. You cannot get leprosy from a casual contact with a person who has Hansen's disease like:

- Shaking hands or hugging
- Sitting next to each other on the bus
- Sitting together at a meal

Hansen's disease is also not passed on from a mother to her unborn baby during pregnancy and it is also not spread through sexual contact.

Due to the slow-growing nature of the bacteria and the long time it takes to develop signs of the disease, it is often very difficult to find the source of infection. The disease has an incubation period (the time between infection and the appearance of the first symptoms) of up to five years. Symptoms may not appear for as long as 20 years. According to the New England Journal of Medicine, an armadillo native to the southern United States can also carry and transmit the disease to humans. Cases of tuberculoid leprosy transmitted through tattooing have been reported, mainly in India.

8.7. SYMPTOMS

Symptoms mainly affect the skin, nerves, and mucous membranes.

The disease can cause skin symptoms such as:

- Discolored patches of skin, usually flat, that may be numb and look faded (lighter than the skin around)
- Growths (nodules) on the skin
- Thick, stiff or dry skin
- Painless ulcers on the soles of feet
- Painless swelling or lumps on the face or earlobes
- Loss of eyebrows or eyelashes

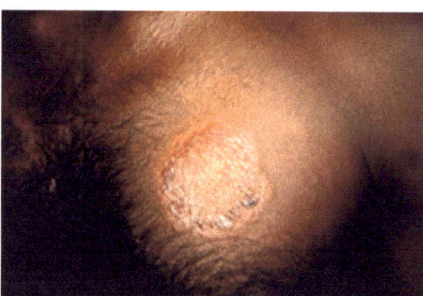

Figure 8.3

A large, discolored lesion on the chest of a person with Hansen's disease.
Source: (https://www.cdc.gov/leprosy/images/health-care-workers/healthcare-1.jpg)

8.7.1. Symptoms caused by damage to the nerves are:

- Numbness of affected areas of the skin
- Muscle weakness or paralysis (especially in the hands and feet)
- Enlarged nerves (especially those around the elbow and knee and in the sides of the neck)

- Eye problems that may lead to blindness (when facial nerves are affected)
- Enlarged nerves below the skin and dark reddish skin patch overlying the nerves affected by the bacteria on the chest of a patient with Hansen's disease. This skin patch was numb when touched.

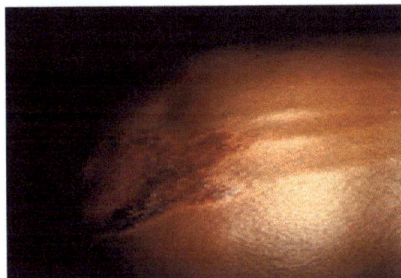

Figure 8.4

Enlarged nerves below the skin and dark reddish skin patch overlying the nerves affected by the bacteria on the chest of a patient with Hansen's disease. This skin patch was numb when touched.
Source: (https://www.cdc.gov/leprosy/images/health-care-workers/healthcare 3.jpg)

8.7.2. Symptoms caused by the disease in the mucous membranes are:

- A stuffy nose
- Nose bleeds
- Since Hansen's disease affects the nerves, loss of feeling or sensation can occur. When loss of sensation occurs, injuries such as burns may go unnoticed. Because you may not feel the pain that can warn you of harm to your body, take extra caution to ensure the affected parts of your body are not injured.
- If left untreated, the signs of advanced leprosy can include:
- Paralysis and crippling of hands and feet
- Shortening of toes and fingers due to reabsorption
- Chronic non-healing ulcers on the bottoms of the feet
- Blindness
- Loss of eyebrows
- Nose disfigurement

8.8. COMPLICATIONS

- Delayed diagnosis and treatment can lead to serious complications. These can include:
- Permanent nerve damage, Hair loss particularly on the eyebrows and eyelashes, Blindness, Chronic nasal congestion, nosebleeds, and collapse of the nasal septum,

sensory loss in the hands and feet, Erectile dysfunction and infertility, Kidney failure, muscular weakness and disfigurement of the physique.

8.9. DIAGNOSIS

Depending on the form of leprosy suspected by the physician, the following specimens may be collected:

- Skin smears from the earlobes, elbows, and knees
- Skin biopsy from edges of active patches
- Nerve biopsy from thickened nerves

8.9.1. SKIN BIOPSY

In the multibacillary form of disease, tissue biopsy of various affected sites may reveal typical histopathologic changes that show large numbers of foam cells. Foam cells are macrophages that have ingested, or phagocytized, M. leprae bacteria, but are unable to digest the organisms, who in turn multiply and use the macrophage as a method of transport throughout the body. This is how the bacteria cause the multiple lesions that may appear in all parts of the patient's body.

8.9.2. SMEAR TEST

A smear can be obtained from nasal mucosa, an ear lobe, and/or skin lesions. Ziehl-Neelsen stain is used to visualize the presence mycobacterium.

8.9.3. SEROLOGY

Currently, diagnosis can be based on the phenolic glycolipid 1 (PGL-1) antibody titer which is useful in multibacillary cases and on polymerase chain reaction (PCR) detection of the bacillus is highly specific and sensitive.

8.10. PREVENTION

The best way to prevent leprosy is to avoid long-term, close contact with an untreated, infected person. Early case detection and proper treatment with MDT is the main strategy to prevent impairments and disabilities among leprosy patients. The health worker should also give appropriate advice on selfcare at home by the patient and his or her family members. It is important to understand that a leprosy patient must complete a full course of MDT.

8.11. TREATMENT

In 1981 the WHO introduced multidrug therapy (MDT) with rifampicin, clofazimine, and dapsone (diamin- odiphenyl sulfone) for first-line treatment. All patients should receive this drug combination monthly under super-vision. Studies show that MDT is highly effective against leprosy and has minimal side-effects. Minocycline, ofloxacin, and clarithromycin are among the drugs used as second-line treatments. In 1992, the vaccine Bacillus Calmette-Guérin (BCG) combined with M leprae was introduced which is very effective against Leprosy and every child should immunize with BCG.

REFERENCES

- Bhattacharya S, Vijayalakshmi N, Parija SC (1 October 2002). "Uncultivable bacteria: Implications and recent trends towards identification". Indian journal of medical microbiology. 20 (4): 174–7.
- Brosch, Roland; Stinear, Timothy P. (11 November 2016). "Leprosy in red squirrels". Science. 354 (6313): 702–703.
- Ishida Y, Pecorini L, Guglielmelli E; Pecorini l; Guglielmelli e (July 2000). "Three cases of pure neuritic (PN) leprosy at detection in which skin lesions became visible during their course". Nihon Hansenbyo Gakkai Zasshi. 69 (2): 101–6.
- Jardim MR, Antunes SL, Santos AR, Nascimento OJ, Nery JA, Sales AM, Illarramendi X, Dupprc N, Chimelli L, Sampaio EP, Sarno EP; Antunes; Santos; et al. (July 2003). "Criteria for diagnosis of pure neural leprosy". J. Neurol. 250 (7): 806–9.
- Mendiratta V, Khan A, Jain A; Khan; Jain (2006). "Primary neuritic leprosy: a reappraisal at a tertiary care hospital". Indian J Lepr. 78 (3): 261–7.
- Meredith, Anna; Del Pozo, Jorge; Smith, Sionagh; Milne, Elspeth; Stevenson, Karen; McLuckie, Joyce (September 2014). "Leprosy in red squirrels in Scotland". Veterinary Record. 175 (11): 285–286.
- Mishra B, Mukherjee A, Girdhar A, Husain S, Malaviya GN, Girdhar BK; Mukherjee; Girdhar; Husain; Malaviya; Girdhar (1995). "Neuritic leprosy: further progression and significance". Acta Leprol. 9 (4): 187–94.
- Ridley DS, Jopling WH; Jopling (1966). "Classification of leprosy according to immunity. A five-group system". Int. J. Lepr. Other Mycobact. Dis. 34 (3): 255–73.
- Rodrigues LC, Lockwood DNj; Lockwood (June 2011). "Leprosy now: epidemiology, progress, challenges, and research gaps". Lancet Infect Dis. 11 (6): 464–70.
- Rodrigues LC; Lockwood DNj (June 2011). "Leprosy now: epidemiology, progress, challenges, and research gaps.". The Lancet infectious diseases. 11 (6): 464–70.
- Schreuder, P.A.M.; Noto, S.; Richardus J.H. (January 2016). "Epidemiologic trends of leprosy for the 21st century". Clinics in Dermatology. 34 (1): 24–31.
- Davey TF, Rees RJ. The nasal discharge in leprosy: clinical and bacteriological aspects.   Lepr Rev. 1974; 25:121---34.
- Ghorpade A. Ornamental tattoos and skin lesions. Tattoo inoculation borderline tuberculoid leprosy. Int J Dermatol. 2009; 48:11---3.

- Centers for Disease Control and Prevention (CDC), (https://www.cdc.gov/leprosy/)
- WHO (2016) World Leprosy report, (http://www.who.int/mediacentre/factsheets/fs101/en/)
- Eichelmann K., González González S.E., Salas-Alanis J.C., Ocampo- Candiani J. Leprosy. An Update: Definition, Pathogenesis, Classification, Diagnosis, and Treatment. Actas dermosifiliogr. 2013;104(7):554---563

9. TUBERCULOSIS

Simani Mohanty
T. K. Barik

9.1. INTRODUCTION

Tuberculosis is an infectious disease caused by the bacillus *Mycobacterium tuberculosis* or *Mycobacterium bovis*. Both are gram-positive, acid and alcohol fast, aerobic, non-spore forming rods, classified under the actinomycetes. It typically affects the lungs (pulmonary TB) but can also affect other sites (extrapulmonary TB). The disease is spread when people who are sick with pulmonary TB expel bacteria into the air, for example by coughing. Overall, a relatively small proportion (5–15%) of the estimated 2–3 billion people infected with *M. tuberculosis* will develop TB disease during their lifetime. However, the probability of developing TB disease is much higher among people infected with HIV.

Tuberculosis infection is characterized by a complex immunologic response, which leads to a unique host-pathogen interaction therefore make it difficult to treat and control. Moreover TB is a poverty related disease and has severe social implications.

9.2. GEOGRAPHIC DISTRIBUTION

Tuberculosis is a worldwide infection from which some three million people die annually. The countries with the largest number of patients with tuberculosis are Bangladesh, Brazil, China, India, Indonesia, Nigeria, Pakistan, the Philippines, and Vietnam. The rate of disease is highest in sub-Saharan Africa. In developing countries tuberculosis accounts for 26% of avoidable deaths.

9.3. ORIGIN AND HISTORY

Tuberculosis is one of the most ancient diseases. Its history dates back to 5000 B.C. of Neolithic period. In Vedas, it is called as "Rajayakshma" (meaning "wasting disease"). It has been referred to in the Vedas and Ayurvedic Samhitas.

"Consumption" and "Phthisis" were terms historically used to describe TB, whichwas responsible for one in four deaths in the 19th century. The word "tubercle" was first used in the seventeenth century by a Dutchman, Franciscus Silvius, of Leyden, to describe the lung lesions. Later (1839), Johann Schönlein called the disease "tuberculosis." It was not until 1882 that Robert Koch identified and described Mycobacterium tuberculosis. The intracellular pathogen that causes tuberculosis, was discovered in 1882 by Robert Koch. In India, the first open air sanatorium for treatment and isolation of TB patients was founded in 1906 in Tiluania, near Ajmer, followed by one in Almora two years later.

9.4. CAUSATIVE AGENT AND CNS TUBERCULOSIS

It is an aerobic, non-motile, nonspore-forming, and acid-fast bacillus that infects primarily humans. Infection occurs through inhalation of droplet nuclei containing the bacilli, leading to deposition in the lung alveoli, interact with alveolar macrophages through a multitude of different receptors then spreads to the draining lymph nodes forming primary Ghon's complex, there exists a low-level but significant bacteremia, in which *M. tuberculosis* disseminates to distant sites of the body that are highly oxygenated, including the brain.

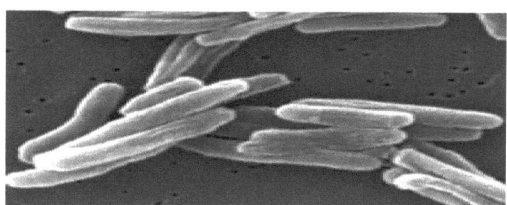

Figure 9.1

Mycobacterium tuberculosis
Source: Wikipedia

In CNS, the disease begins with the development of small TB foci called Rich foci in the brain, spinal cord, or meninges, rupture of a Rich focus into the subarachnoid space heralds the onset of meningitis and less commonly as tubercular encephalitis, intracranial tuberculoma, or a tuberculous brain abscess. Tumor necrosis factor alpha (TNF-α) is essential for breaking the blood–brain barrier, microglial cells are infected and are the principal target in the CNS. Production of γ-interferon from T-lymphocytes specific for mycobacterial peptides, activates microglia enabling more efficient intracellular killing of tubercle bacilli. *M. tuberculosis*-infected microglia also produce robust amounts of several cytokines and chemokines in vitro, including TNF-α, interleukin-6 (IL-6), IL-1 β, CCL2, CCL5, and CXCL10. Subsequent complications are related to adhesion formation, an obliterative vasculitis, and an encephalitis or myelitis.

9.5. SYMPTOMS

TB bacteria most commonly grow in the lungs, and can cause symptoms such as:
- A bad cough that lasts 3 weeks or longer
- Pain in the chest
- Coughing up blood or sputum (mucus from deep inside the lungs)
- Other symptoms of TB disease may include:
- Weakness or fatigue
- Weight loss
- No appetite

- Chills
- Fever
- Sweating at night

9.6. LIFE CYCLE

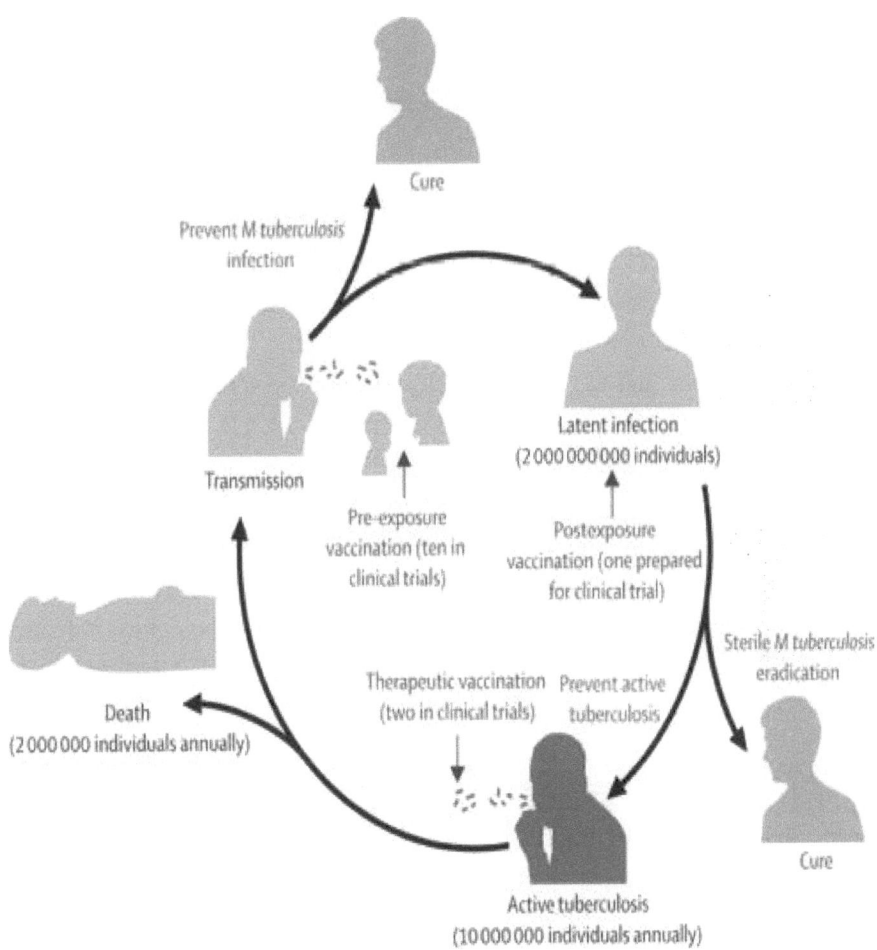

Figure 9.2

Source: http://tbtuberculosis0.weebly.com/mycobacterium-life-cycle.html

9.7. TRANSMISSION

The bacteria that cause TB, is spread through the air from person to person. People nearby may breathe in these bacteria and become infected. There are two types of TB conditions: latent TB infection and TB disease. About one-third of the world's population has latent TB, which means people have been infected by TB bacteria but are not (yet) ill with the disease and cannot transmit the disease. If TB bacteria become active in the body and multiply, the person will go from having latent TB infection to being sick with TB disease.

In most cases, infection is transmitted by inhalation of *M. tuberculosis* containing microscopic droplets originating from cases of active pulmonary tuberculosis. M. tuberculosis is carried in airborne particles called droplet nuclei that can be generated when persons who have pulmonary or laryngeal TB disease cough, sneeze, shout, or sing. The particles are approximately 1–5 μm; normal air currents can keep them airborne for prolonged periods and spread them throughout a room or building. *M. tuberculosis* is usually transmitted only through air, not by surface contact. After the droplet nuclei are in the alveoli, local infection might be established, followed by dissemination to draining lymphatics and hematogenous spread throughout the body. Infection occurs when a susceptible person inhales droplet nuclei containing *M. tuberculosis*, and the droplet nuclei traverse the mouth or nasal passages, upper respiratory tract, and bronchi to reach the alveoli. Persons with TB pleural effusions might also have concurrent unsuspected pulmonary or laryngeal TB disease.

Usually within 2–12 weeks after initial infection with *M. tuberculosis*, the immune response limits additional multiplication of the tubercle bacilli, and immunologic test results for *M. tuberculosis* infection become positive. However, certain bacilli remain in the body and are viable for multiple years. This condition is referred to as latent tuberculosis infection (LTBI). Persons with LTBI are asymptomatic (they have no symptoms of TB disease) and are not infectious.

9.8. TREATMENT

Treatment includes first line drugs followed by second line drug system if not curable. TB Antibiotics have been around since 1944 Following streptomycin, p-aminosalicylic acid (1949), isoniazid (1952), pyrazinamide (1954), cycloserine (1955), ethambutol (1962), and rifampin (rifampicin; 1963) were introduced as anti-TB agents. Aminoglycosides such as capreomycin, viomycin, kanamycin, and amikacin, and the newer quinolones (e.g. moxifloxacin, levofloxin, ofloxacin, and ciprofloxacin) are effective but usually are used in drug resistance situations.

Drug resistant (DR) Tuberculosis is defined as a strain of Tuberculosis that is resistant to one of the "first line" drugs listed above. When drug-resistance strains pop up many treatment options are suppressed and doctors are forced to use more dangerous second-line drugs. The most prevalent resistance strains render Isomiazid and Rifampicin useless and Rifabutin will only work 30% of time if the strain is resistant to Rifampicin because of molecular similarity.

Multi- Drug resistant (MDR) Tuberculosis is defined as a strain of Tuberculosis that is resistant to more than one first line drug at least isoniazid and rifampicin. Globally, about 4.5% of new tuberculosis (TB) cases are multi-drug-resistant (MDR), i.e. resistant to the two most powerful first-line anti-TB drugs. The MDR *Mycobacterium tuberculosis* M family, which belongs to the Haarlem lineage, is highly prosperous in Argentina and capable of building up further drug resistance without impairing its ability to spread. The duration also increases from 6-8 months to 18-24 months with MDR strains.

If the microorganism is resistant to the standard drugs, then it will be necessary to administer more toxic medications such as

- Ethionamide
- Cycloserine
- Capreomycin
- Kanamycin
- Streptomycin
- Fluoroquinolones

DOTS (directly observed treatment, short course) is the internationally recommended control strategy for TB. This strategy includes the delivery of a standard short course of drugs, lasting 6 months for new patients and 8 months for retreatment patients, to individuals diagnosed with TB. The delivery includes the direct observation of therapy (DOT), either by a health worker or by someone nominated by the health worker and the patient for this purpose (sometimes called a DOT supporter). The strategy has been promoted widely and implemented globally.

DOTS or Directly Observed Treatment Short course is the internationally recommended strategy for TB control that has been recognized as a highly efficient and cost-effective strategy. DOTS comprises five components.

1. Sustained political and financial committment. TB can be cured and the epidemic reversed if adequate resources and administrative support for TB control are provided.

2. Diagnosis by quality ensured sputum-smear microscopy. Chest symptomatics examined this way helps to reliably find infectious patients.

3. Standardized short-course anti-TB treatment (SCC) given under direct and supportive observation (DOT). Helps to ensure the right drugs are taken at the right time for the full duration of treatment.

4. A regular, uninterrupted supply of high quality anti-TB drugs. Ensures that a credible national TB programme does not have to turn anyone away.

5. Standardized recording and reporting. Helps to keep track of each individual patient and to monitor overall programme performance.

9.9. TESTING FOR PULMONARY TB

Any person who has signs and symptoms suggestive of TB including a cough for more than 2 weeks, and a fever for more than 2 weeks, significant weight loss, haemoptysis (coughing blood) etc. and any abnormality in a chest radiograph should be evaluated to find out if the have TB.

Children with a persistent fever and/or cough for more than 2 weeks, children who have a loss of weight or no weight gain, and/or children who are household contacts of people who have already been diagnosed as having pulmonary TB must be evaluated for TB.

9.9.1. SCREENING FOR TB

People living with HIV (PLHIV), people who are malnourished, who have diabetes or cancer, and people on steroid therapy should be regularly screened for signs and symptoms suggestive of TB. Enhanced case finding should be undertaken in certain "high risk" populations such as healthcare workers, prisoners, slum dwellers. There should also be enhanced case finding in certain occupational groups such as mineworkers, as in some countries such as South Africa, there is known to be a high level of TB among miners.

Enhanced case finding means having a high level of suspicion for TB in all encounters. Then excluding TB (or indeed identifying TB) using a combination of clinical queries, radiographic and microbiologic testing.

9.10. DIAGNOSIS

There are several tests available for the diagnosis of TB. Sputum smear microscopy is the most common TB test, and the majority of people with TB are sputum smear positive.

9.11. DIAGNOSIS OF TB IN INDIA

There are a number of diagnostic TB tests currently available. These include:

The TB skin test

The TB skin test is a widely used test for diagnosing TB. In countries with low rates of TB it is often used to test for latent TB infection. The TB skin test involves injecting a small amount of fluid (called tuberculin) into the skin in the lower part of the arm. Then the person must return after 48 to 72 hours to have a trained health care worker look at their arm. The health care worker will look for a raised hard area or swelling, and if there is one then they will measure its size. They will not include any general area of redness. The TB skin test result depends on the size of the raised hard area or swelling. The larger the size of the affected area the greater the likelihood that the person has been infected with TB bacteria at some time in the past. But interpreting the TB skin test result that is whether it is a positive result, may also involve considering the lifestyle factors of the person being tested for TB. The TB skin test also cannot tell if the person has latent TB or active TB disease. False positive results happen with the TB skin test because the person has been infected with a different type of bacteria, rather than the one that causes TB.

9.12. FLUORESCENT MICROSCOPY

The use of fluorescent microscopy is a way of making sputum TB tests more accurate. With a fluorescent microscope the smear is illuminated with a quartz halogen or high pressure mercury vapour lamp, allowing a much larger area of the smear to be seen and resulting in more rapid examination of the specimen.

One disadvantage though is that a mercury vapour lamp is expensive and lasts a very short time. Such lamps also take a while to warm up, they burn significant amounts of electricity, and electricity supply problems can significantly shorten their life span.

Chest X-ray as a TB test

Where available chest X-ray should be used as a screening tool.

If a person has had TB bacteria which have caused inflammation in the lungs, an abnormal shadow may be visible on a chest x-ray. Also, acute pulmonary TB can be easily seen on an X-ray. However, what it shows is not specific. A normal chest X-ray cannot exclude extra pulmonary TB.

Also, in countries where resources are more limited, there is often a lack of X-ray facilities.

Cartridge Based Nucleic Acid Amplification Test (CB NAAT)

The CB NAAT is known as the Gene Xpert in most countries other than India. This is the preferred first diagnostic test in children and people with TB and HIV co-infection.

TB Interferon gamma release assays (IGRAs)

The Interferon Gamma Release Assays (IGRAs), are a new type of more accurate TB test. In this context referring to an assay is simply a way of referring to a test or procedure.

IGRAs are blood tests that measure a person's immune response to the bacteria that cause TB. The immune system produces some special molecules called cytokines. These TB tests work by detecting a cytokine called the interferon gamma cytokine. In practice you carry out one of these TB tests by taking a blood sample and mixing it with special substances to identify if the cytokine is present.

9.13. PREVENTION

Some other steps toward preventing the spread of TB include:

Improving the ventilation in indoor spaces so there are fewer bacteria in the air.

Using germicidal ultraviolet lamps to kill airborne bacteria in buildings where people at high risk of tuberculosis live or congregate.

Treating latent infection before it becomes active.

Using directly observed therapy (DOT) in people with diagnosed tuberculosis (latent or active) to raise the likelihood of the disease being cured.

A vaccine for tuberculosis called *bacilli Calmette- guerin* or BCG is used in parts of the world with high rates of the infection to prevent serious complications such as meningitis.

Early diagnosis and treatment is the most effective way to prevent the spread of tuberculosis.

A person with infectious tuberculosis can infect up to 10–15 other people per year. But once diagnosed with TB, and started on treatment, the majority of patients are no longer infectious after just two weeks of taking the medication.

The first part of TB prevention is to stop the transmission of TB from one adult to another. This is done through firstly, identifying people with active TB, and then curing them through the provision of drug treatment. With proper TB treatment someone with TB will very quickly not be infectious and so can no longer spread the disease to others. The second main part of TB prevention is to prevent people with latent TB from developing active, and infectious, TB disease.

In 1982, a century after Dr. Koch's announcement, the first World TB Day was sponsored by the World Health Organization (WHO) and the International Union Against Tuberculosis and Lung Disease (IUATLD). The event was intended to educate the public about the devastating health and economic consequences of TB, its effect on developing countries, and its continued tragic impact on global health. It is a valuable opportunity to educate the public about the devastation TB can spread and how it can be stopped. Superstition and religion also played large roles in the confusion surrounding tuberculosis and how to combat it, leading to some semi-effective techniques and many ineffective or even harmful ones. For much of human history, tuberculosis wasn't so much treated as contained or at best battled to a draw. The advent of modern medicine changed this, but overall tuberculosis' incidence and mortality rates have gone thorough peaks and troughs influenced as much by environmental factors as treatment strategies. Antimicrobial chemotherapy is the term used to describe drug treatment aimed at reducing the population of a particular microbe (in this case *Mycobacterium tuberculosis*). Today, there are a number of methods for treating tuberculosis, including potential new vaccines on the horizon. This is a massive improvement considering the paucity of options available for much of human history. However, new challenges have developed with growing incidences of drug resistant tuberculosis, including MDR (multiple drug resistant) and XDR (extensively drug resistant) strains. XDR strains are a serious threat, especially for immunodeficient patients who are significantly less likely to recover without the assistance of effective drugs. Luckily, there is a global effort to control and eventually eradicate tuberculosis.

REFERENCES

- National Strategic Plan (2012-2017) for Tuberculosis – Directorate of Health Services, Central TB Division, Ministry of Health & Family Welfare (MoHFW) Government of India, New Delhi, www.tbcindia.nic.in - See more at: http://www.tbfacts.org/tb-testing-diagnosis-india/#sthash.S9kjz349.dpuf
- Centers for Disease Control and Prevention(CDC)
- WHO (2017) Fact Sheet: http://www.who.int/mediacentre/factsheets/fs104/en/
- Centers for Disease Control and Prevention(CDC),
- https://www.cdc.gov/tb/topic/basics/default.htm
- Simona Luca; Traian Mihaescu. History of BCG Vaccine. Maedica – a Journal of Clinical Medicine 2013; 8(1): 53-58
- Thomas M. Daniel. The history of tuberculosis. Respiratory Medicine 2006 100, 1862–1870. doi:10.1016/j.rmed.2006.08.006.

10. TYPHOID

Usha Rani Acharya
Bibarani Tripathy
Surya Narayan Swain
Bijayalaxmi Sahu

10.1. INTRODUCTION

Typhoid is a systemic febrile illness caused by *Salmonella enterica* serotype Typhi (formerly named *Salmonella typhi*). This highly adapted, human-specific pathogen has evolved remarkable mechanisms for persistence in its host that help to ensure its survival and transmission. According to global estimates, there are approximately 27 million cases of enteric fever and over 200,000 deaths each year caused by typhoid fever. *S. typhi*, was isolated by Gaffkey in Germany in 1884. Typhoid is usually spread by ingestion of food or water contaminated by urinary or fecal matters containing *S. enterica* serotype *typhi*. It is a sporadic disease in developed countries that occurs mainly in returning travellers. In endemic areas, identified risk factors for disease include eating food prepared outside the home, drinking contaminated water, having a close contact or relative with recent typhoid fever and poor housing with inadequate facilities for hygiene.

10.2. OCCURENCE

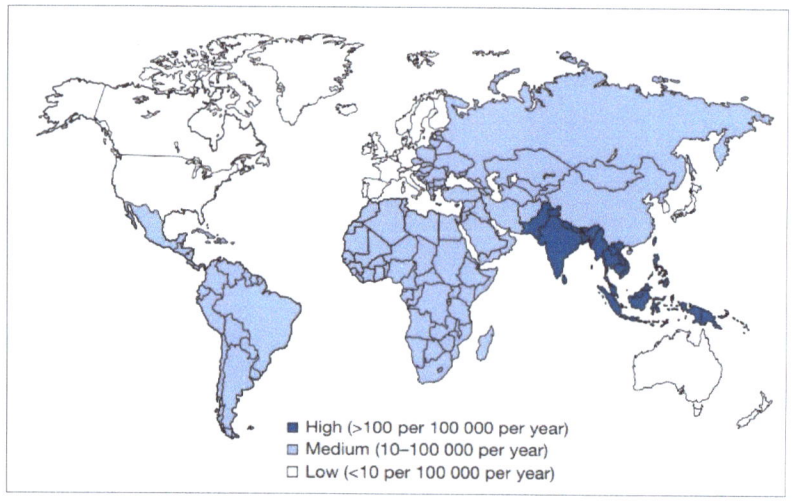

Figure 10.1

Global Distribution of Typhoid fever.
Source: Hunter's Tropical Medicine and Emerging Infectious Disease (9th Ed.)

In general, an estimated 22 million cases of typhoid and 200,000 to 600,000 deaths occur annually worldwide. Regions with a high incidence of typhoid fever (>100/100,000 cases/year) include south-central Asia and Southeast Asia. Regions of medium incidence (10-100/100,000 cases/year) include the rest of Asia, Africa, Latin America and the Caribbean, and Oceania, except for Australia and New Zealand. The Indian sub-continent (India, Pakistan, and Bangladesh) accounts for the majority of all reported cases of typhoid that are a result of travel to endemic areas. In endemic regions, typhoid fever is more common in urban than rural areas and among young children and adolescents (aged 1 to 15 years) (Mandell et. Al, 2010).

10.3. EPIDEMIOLOGY

Typhoid continues to be a global health problem infected by *Salmonella typhi*. Humans are the only natural host and reservoir. Up to 5% of patients continue to harbour *S. Typhi* in their intestinal tract and gallbladder for months or years ("asymptomatic carriers"). Most commonly, food or waterborne transmission occurs as a result of faecal contamination by ill or asymptomatic chronic carriers. The incubation period is usually 8–14 days, but may range from 3 days up to 2 months. Some 2–5% of infected people become chronic carriers who harbour S. typhi in the gall bladder. Chronic carriers are greatly involved in the spread of the disease. Important vehicles in some countries include shellfish (particular lyoysters) from sewage-contaminated beds, fruit and vegetables eaten raw, and milk/milk products contaminated through hands of carriers. *S. typhi* can be transmitted sexually, including by anal and oral sex. Hence, Patients infected with HIV are at significantly increased risk of severe disease transmission due to *S. typhi* and *S. paratyphi*.

10.4. CAUSATIVE AGENT

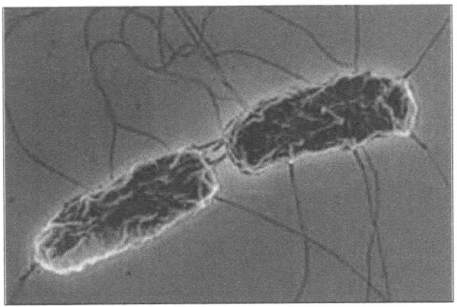

Figure 10.2

Image: *Salmonella enterica* serovar Typhi.

Source: microbewiki.kenyon.edu

Salmonella enterica serotype *typhi* is a Gram-negative, rod-shaped, facultative, anaerobic bacteria, member of the family Enterobacteriaciae. A very similar but often less severe disease is caused by *Salmonella serotype paratyphi* A. The genus salmonella comprises of two species: *Salmonella typhi* and *Salmonella bongori*. Some of the serovars of *S. enterica* are host specialists that infect only humans, whereas others are host generalists that occur in humans and many other mammalian species. *Salmonella typhi* is a human-restricted pathogen. It is still controversial as to why this pathogen does not infect other organisms and has such a selective host behaviour. Like many infectious diseases, typhoid fever is the manifestation of the outcome of a complex cross talk between the human host, its environment and the microbe, with many acquired, random and genetic factors coming into play. The severity of typhoid depends on:

- genetic properties of the pathogen
- the bacterial inoculum that effectively reaches its site of entry into the body
- specific resistance mechanisms of the host

A characteristic phenomenon of host restriction such as that found for *S. typhi* is gene loss. In Typhi strain there are about 204 inactivated pseudogenes, which may explain its host restriction to humans. *S. typhi* has several unique features, the genetic basis of many of which is known as a result of early genetic studies and the recent sequencing of the whole genome. The bacterium is serologically posistive for lipopolysaccharide antigens O-9 and O-12, protein flagellar antigen Hd, and polysacharide capsular antigen Vi. The Vi capsular antigen is largely restricted to *S. typhi*. Gene clusters, present inside the genome of *S. typhi*, are unique to particular bacteria in specific environments or may contribute to pathogenicity. *S. typhi* has several large insertions in its genome, termed salmonella pathogenicity islands, that are thought to be recent horizontal acquisitions and that encode genes important for survival in the host.

10.5. MORPHOLOGY AND STAINING

Salmonella typhi is a short rod, that measure about 0.5-0.8µ in width and 1-3.5µ in length. Generally, It is actively motile, non sporogenous and found in non-encapsulated form. It is stained with aniline dyes and are gram -ve. *S. typhi* can directly be isolated from blood. Bile facilitates the growth of the pathogen. In fact, bile is not growth promoting but apparently reduces the bactericidal activity of blood. It remains alive in cultures for months even years, if moisture is supplied. Their survival in ice or snow for 3 months accounts for the water borne epidemics.

10.6. MULTIPLICATION AND PROPAGATION

During an acute infection, *S.typhi* multiplies in mononuclear phagocytic cells before being released into the bloodstream. When this bacteria first enters the human body it initially propagates inside the intestinal tract and spreads throughout the peripheral lymphatic system, such as the bone marrow and Peyer's patches, to cause typhoid fever. The bacteria can invade the intestinal mucosa potentially through microfold cells and establishs a clinically undetectable infection involving significant systemic dissemination and a transient primary bacteremia. The bacteremia rapidly penetrate the mucosal epithelium via either microfold cells and arrive in the lamina propia, where they rapidly penetrate elicit an influx of macrophages that ingest the bacili but do not generally kill them. After infection, the incubation period may not always be followed by clinical symptoms. Those who go on to develop typhoid become fatigued, and the fever begins to rise in a classical stepwise manner. The most common sites of secondary infection are the liver, spleen, bone marrow, gallbladder, and Peyer's patches of the terminal ileum.

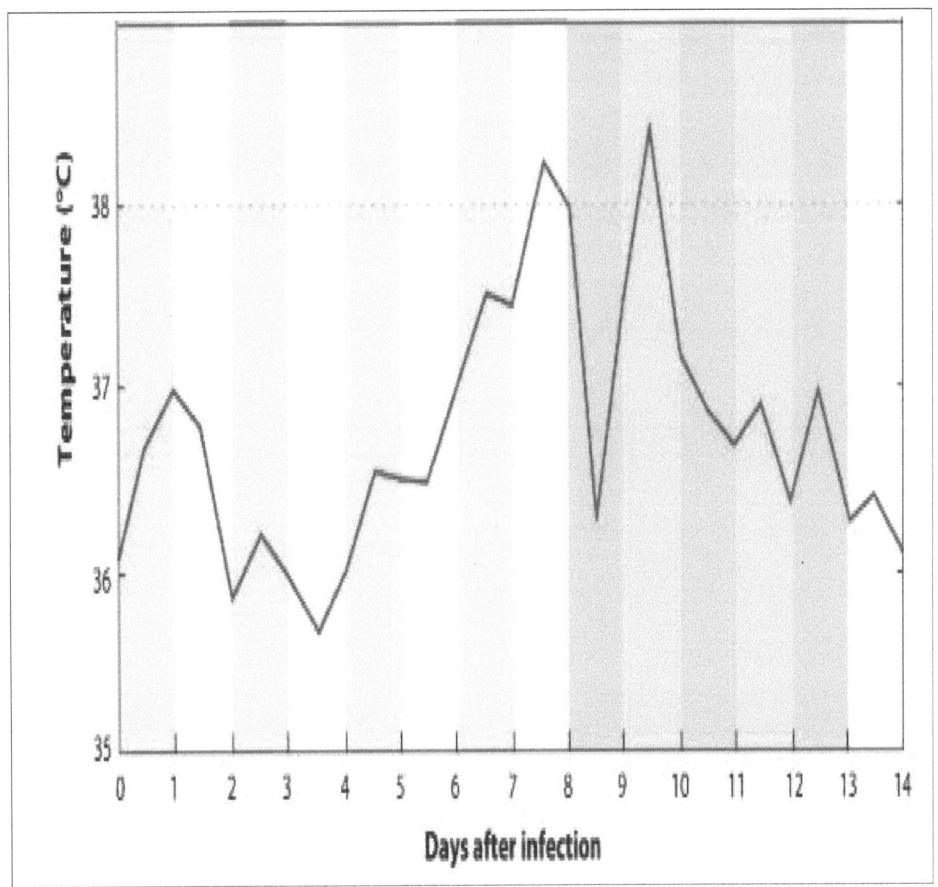

Figure 10.3

Variations of body temperature during post-infection period
Source: www.microbewiki.kenyon.edu/Salmonella enterica serovar typhi.

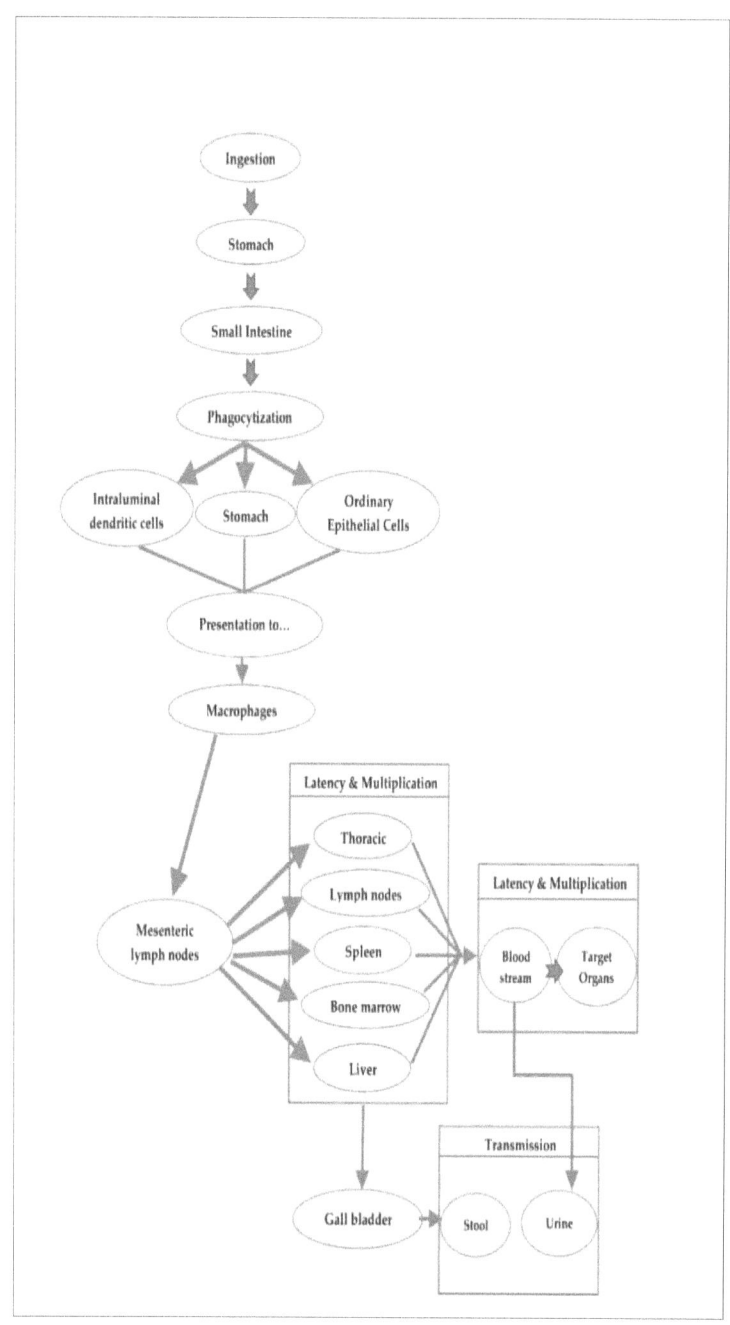

Figure 10.4

Infection pathway *of Salmonella enterica serovar typhi*.
Source: www.microbewiki.kenyon.edu/*Salmonella enterica serovar typhi*.

10.7. SYMPTOMS

The clinical presentation of typhoid fever varies from a mild illness with low-grade fever, malaise, and slightly dry cough to a severe clinical picture with abdominal discomfort and multiple complications. Many factors influence the severity and overall clinical outcome of the infection. They include:

- The duration of illness before the initiation of appropriate therapy.
- The choice of antimicrobial treatment.
- Age
- The virulence of the bacterial strain.
- The quantity of inoculum ingested etc.
- Acute non-complicated fever

Acute typhoid fever is characterized by prolonged fever, disturbances of bowel function (constipation in adults, diarrhoea in children), headache, malaise and anorexia. Bronchitis cough is common in the early stage of the illness. During the period of fever, up to 25% of patients show exanthema (rose spots), on the chest, abdomen and back.

10.8. COMPLICATED FEVER

Acute typhoid fever may be severe, with up to 10% of patients developing serious complications. Intestinal perforation is also occasionally reported in about 3% of hospitalized cases. Abdominal discomfort, the symptoms and signs of intestinal perforation and peritonitis, are among other complications. Altered mental status in typhoid patients has been associated with a high case-fatality rate. Such patients generally have delirium and may progress to coma.

The hallmark of typhoid fever is a prolonged fever that may persist up to 4 to 8 weeks in untreated cases. Even though the illness may be mild and brief, in rare cases an acute severe infection progresses into multiple organ failure, disseminated intravascular coagulation and deleterious effects in central nervous system resulting in early death. In other instances, intestinal bleeding and perforation of the necrotic Peyer's patches can occur in the third or fourth week of illness, and result in late death. Other gastrointestinal manifestations include constipation rather than diarrhoea and often accompanied by abdominal tenderness.

10.9. CONTAMINATION AND TRANSMISSION

The infection is transmitted by ingestion of food or water contaminated with faeces. Ice cream is recognized as a significant risk factor for the transmission of typhoid fever. Shellfish taken from contaminated water, and raw fruit and vegetables grown in contaminated soil, have been sources of immediate outbreaks. The highest incidence occurs where water supplies serving large populations are contaminated with faeces. Epidemiological data suggest that waterborne transmission of *S. typhi* usually involves small inocula, whereas food borne transmission is associated with large inocula and high attack rates over short periods. The inoculum size and the type of vehicle in which the organisms are ingested greatly influence both the attack rate and the incubation period.

10.10. DIAGNOSIS

The definitive diagnosis of typhoid fever depends on the isolation of *S. typhi* from blood, bone marrow or a specific anatomical lesion. Blood cultures may be more sensitive than stool or urine cultures in the first week of disease. *S. typhi* can also be cultured from bone marrow; in fact, this is the most sensitive method of isolating of S. typhi. Serologic tests based on agglutinating antibodies (Widal test) are generally of little diagnostic value. Newer, rapid serodiagnostic tests using enzyme-linked immunosorbent assay (ELISA) and dipstick techniques have better sensitivity and specificity than the Widal test and may be useful in acute outbreak settings. However, they are not useful for individual patient diagnosis. An ELISA for antibodies to the capsular polysaccharide Vi antigen may be useful in diagnosing chronic carriers of S. typhi.

The Utah Public Health Laboratory (UPHL) accepts stool, urine and blood specimens for culture. Serotyping and Pulsed-Field Gel Electrophoresis (PFGE) are performed on all positive Salmonella cultures. Clinical laboratories must submit all Salmonella isolates to UPHL for confirmation, serotyping and PFGE analysis.

The Centers for Disease Control and Prevention (CDC) provides Whole Genome Sequencing (WGS) of *S. typhi* isolates. WGS analysis may be useful in preventing outbreaks and to identify the source of infection. CDC also provides Vi antibody testing to identify chronic carriers. Utah Department of Health (UDOH) may therefore be contacted for more information.

10.11. TREATMENT

Supportive measures are important in the treatment of typhoid fever, such as oral or intravenous hydration, the use of antipyretics, and appropriate nutrition and blood transfusions if indicated.

The fluoroquinolones are widely regarded as optimal for the treatment of typhoid fever in adults. They are relatively inexpensive, well tolerated and more rapidly and reliably effective than the former first-line drugs, viz. chloramphenicol, ampicillin, amoxicillin and trimethoprim-sulfamethoxazole. The fluoroquinolones attain excellent tissue penetration, kill *S. typhi* in its intracellular stationary stage in monocytes/macrophages and achieve higher active drug sensitivity levels in the gall bladder compared to other drugs. They produce a rapid therapeutic response, i.e. declining fever and other symptoms in three to five days. Ciprofloxacin, ofloxacin, perfloxacin and fleroxacin have also been proved effective. However, in recent years, there have been many reports of reduced susceptibility for ciprofloxacin.

10.12. SURGERY

Surgery is usually indicated in cases of intestinal perforation. Most surgeons prefer simple closure of the perforation with drainage of the peritoneal fluid. Small bowel resection is indicated for patients with multiple perforations. If antibiotic treatment fails to eradicate the hepatobiliary carriage, the gallbladder should be rejected. Cholecystectomy is not always successful in eradicating the carrier state because of persisting hepatic infection.

10.13. PREVENTION

The major routes of transmission of typhoid fever are through drinking water or eating food contaminated with *S. typhi*. Prevention is based on ensuring access to safe water and by promoting safe food handling practices.

- How is it prevented?
- Avoid uncooked foods, including fruit unless it can be peeled
- Avoid untreated and unfiltered water, including ice
- Drink beverages from sealed containers
- Wash hands thoroughly with soap after going to the toilet and before eating or handling food
- Avoid eating from street vendors
- Avoid taking raw food and shellfish.
- Ensure taking food in thoroughly cooked and in hot condition.
- Cover food items to protect from flies.
- Proper solid waste disposal
- Safe water

Typhoid is a waterborne disease and the main preventive measure is to ensure access to safe water. In urban areas, control and treatment of the water supply systems must be strengthened from catchment to consumer. Safe drinking water should be made available to the population through a piped system or from tanker trucks. In rural areas, wells must be checked for pathogens and chemically treated if necessary.

10.14. VACCINATION

The classic heat-inactivated whole cell vaccine is associated with an unacceptably high rate of adverse effects and has been largely withdrawn from public health use. Globally, two vaccines are currently available for potential use in children. An oral, live-attenuated preparation of the Ty21a strain of *S. typhi* has been shown to have good efficacy (range, 67%– 82%) for up to 5 years. The Vi capsular polysaccharide can be used in children aged 2 years or older. It is administered as a single intramuscular dose, with a booster every 2 years, and has a protective efficacy of 70% to 80%. The vaccines are currently recommended for traveling into endemic areas, but a few countries have introduced large-scale vaccination strategies. WHO recommends that the immunization of school-age children be undertaken wherever the control of the disease is a priority.

Infection caused by *S. typhi* remains an important public health problem, particularly in developing countries. Morbidity and mortality attributable to typhoid fever are once again increasing with the emergence and worldwide spread of *S. typhi* strains that are resistant to most previously useful antibiotics. As a consequence, there is renewed interest in understanding the epidemiology, diagnosis and treatment of typhoid fever and some specific aspects of its pathogenesis. More importantly, perhaps, there is increased need to expand various active typhoid vaccines. Public health authorities should now devise ways of using the two currently available improved typhoid vaccines, parenteral Vi polysaccharide and oral Ty21a, in large-scale nursery-based and school-based immunization programmes, and should monitor their public health impact. However, we should realise, "Prevention is better than cure" and hence, public awareness to combat the spread of typhoid should be the prime focus.

REFERENCES

- www.microbewiki.kenyon.edu/*Salmonella enterica serovar typhi*.
- Hunter's Tropical Medicine and Emerging Infectious Disease (9th Ed.)
- Mandell, Douglas and Bennett's (2010). Principles and practice of infectious diseases, 7th edition, pp 2287-2888.
- The diagnosis, treatment and prevention of typhoid fever. World Health Organisation, Guidelines for communicable disease surveillance and response vaccines and biological, 2003.

11. MALARIA

Deepika Panda
T. K. Barik

11.1. INTRODUCTION

Malaria is an ancient mosquito-borne protozoan disease that, worldwide, causes about half a billion cases with one million human deaths every year, mainly in tropical and some sub-tropical countries. Thus, it remains a disease of tremendous public health importance on a global scale. The degree of endemicity varies between countries and even between different areas in the same country. In areas with high transmission of malaria, children under 5 and pregnant women are particularly susceptible to infection, illness and death; more than two thirds (70%) of all malaria deaths occur in this age group. Along with humans malaria also affects birds, lizards, rodents, non-human primates. Technically, in human malaria, the mosquito vector is the "definitive host" because the parasite reproduces in them.

11.2. ORIGIN AND HISTORY

Malaria as a parasitic infectious disease was first identified in 1880. The term malaria originates from Medieval Italian: mala aria - bad air – swampy air was thought to cause it (Close, but no vector). The disease was formerly called ague or marsh fever due to its association with swamps. Probably arose in Africa and evolved with us from primate ancestors. Sir Ronald Ross on 29th August 1897 made a discovery that the malaria parasites are sucked up by female *Anopheles* mosquito and later on injected in the human blood. Hence 29th August is referred as Mosquito Day.

11.3. GEOGRAPHICAL DISTRIBUTION

Malaria can be transmitted throughout urban areas of Africa and to a lesser extent, India. There is usually less risk at altitudes above 1500 m, although in favorable climatic conditions the disease can be observed at altitudes up to almost 3000 m. The risk of infection may also vary according to the season, being highest at the end of, or soon after, the rainy season. In warmer regions closer to the equator malaria transmission is more intense and transmitted year-round. In cooler regions, transmission is less intense and more seasonal. In many temperate areas, such as Western Europe and the United States, economic development and public health measures have succeeded in eliminating malaria. However, most of these areas have *Anopheles* mosquitoes that can transmit malaria, and reintroduction of the disease is a constant risk.

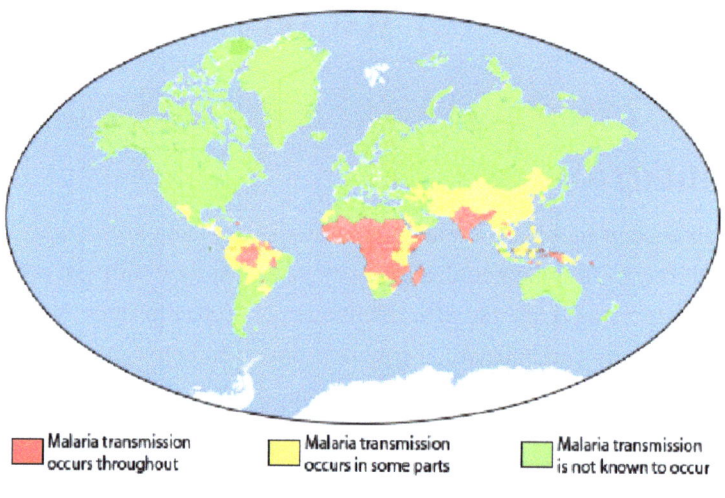

Figure 11.1

Source: Centre for Disease Control

11.4. CAUSATIVE AGENT

There are more than 100 types of *Plasmodium parasites* which can infect a variety of species. *Plasmodium species* are exhibit a heteroxenous life cycle involving a vertebrate host and an arthropod vector. Vertebrate hosts include: reptiles, birds, rodents, monkeys and humans. *Plasmodium species* are generally host and vector specific in that each species will only infect a limited range of hosts and vectors. Scientists have identified five types of *Plasmodium species* that specifically infect humans are:

P. falciparum-It is located worldwide in tropical and suburban areas, but predominately in Africa. About 1 million people are killed by this strain every year globally. The strain can multiply rapidly and can adhere to blood vessel walls in the brain, causing rapid onset of severe malaria including cerebral malaria. Incubation period is 10 days. A very serious result of falciparum infection is blackwater fever, a condition when destruction of patient's erythrocyte occurs and the liberated haemoglobin is excreted in urine.

P. vivax-It has a wide distribution in tropical and temperate zones. Incubation period is 10 days. This strain has a dormant liver stage that can activate and invade the blood after months or years, causing many patients to relapse. It causes benign tertian malaria or vivax malaria fever every 48 hours.

P. ovale-It is located mainly in West Africa, it is biologically and morphologically very similar to *P. vivax*. However, unlike *P. vivax*, this strain can affect individuals who are negative blood group, which is the case for many residents of sub-Saharan Africa. Incubation period is 14 days. It causes mild tertian malaria fever every 48 hours.

P. malariae-It is found in tropical and temperate zones. Incubation period is 27 to 37 days. If left untreated, *P. malariae* can cause a long-lasting, chronic infection that can last a lifetime and which may cause the nephrotic syndrome. It causes quartan malaria fever every 72 hours.

P. knowlesi-It is located in Southeast Asia and associated with macaques (a type of monkey). It can multiply rapidly once a patient is infected, causing an uncomplicated case to become serious very quickly. Fatal cases of infection with this strain have been reported.

11.5. LIFE CYCLE OF PLASMODIUM

The malaria parasite has a complex, multistage life cycle occurring within two living beings, the vector mosquitoes and the vertebrate hosts. The survival and establishment of the parasite within the invertebrate and vertebrate hosts, in intracellular and extracellular environments, is made possible by the presence of more than 5,000 genes and their specialized proteins that help the parasite to invade and grow within multiple cell types and to evade host immune responses (Brain et.al. and Laurence et.al.). The parasite passes through several stages of development such as the sporozoites, merozoites, trophozoites, and gametocytes (sexual stages) and all these stages have their own unique shapes and structures and protein complements. The cell surface proteins and metabolic pathways keep changing during the different stages that help the parasite to evade the host immune system, while also creating problems for the development of drugs and vaccines.

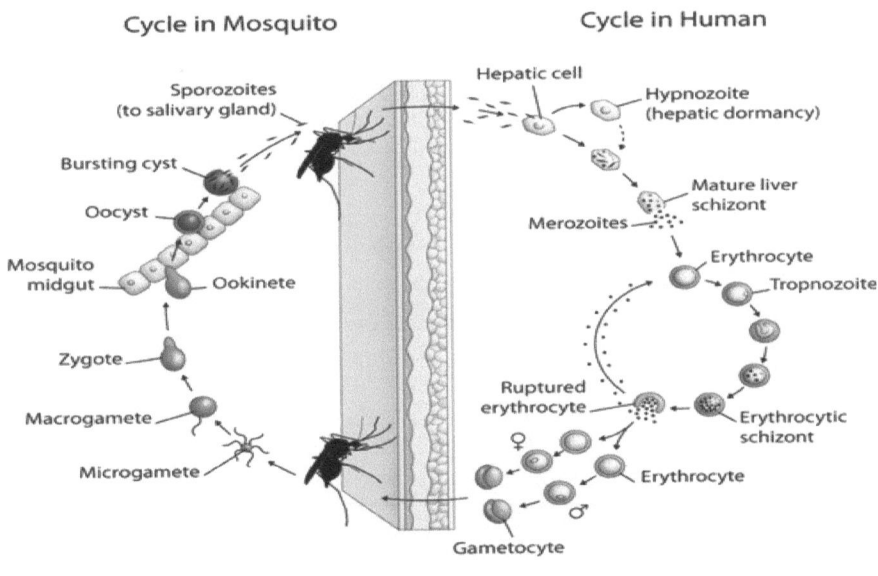

Figure 11.2

Source: "Life cycle of the malaria parasite". Available at: http://ocw.jhsph.edu.

1. The Anopheles mosquito bites a healthy human and inject the *Plasmodium parasite* into the blood stream. At this stage the parasite is in a form known as a sporozoite, which is long and thin and is capable of invading into the cell.
2. The parasite travels in the blood until it reaches the liver where it recognizes and enters into the liver cells and remains for around 10 days. In the liver it undergoes a transformation into thousands of new form of parasites known as a merozoites and finally released into the bloodstream.
3. Each merozoite inside the red blood cell grows and divides asexually to form up to 20 new merozoites. These burst out of the cell and invade neighboring red blood cells. This whole process takes approximately 48 hours.
4. Some parasites do not form merozoites rather develop into a sexual stage of the parasite called gametocytes. These are taken up by a mosquito when they feed on an infected human and transmit to another health individual.
5. Once inside the mosquito gut, the gametocytes change into mature gametes which fuse and develop into an ookinete. The ookinete moves through the lining of the mosquito's gut wall where it forms an oocyst in which thousands of sporozoites are formed. They burst out of the oocyst and travel to the salivary gland of the mosquito where the cycle begins again.

11.6. VECTORS OF MALARIA

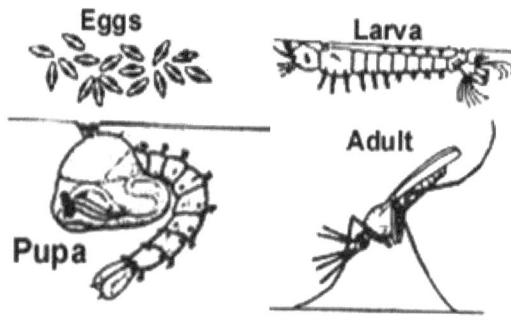

Figure 11.3

Life Cycle of Anopheles mosquito

Source: http://images.aarogya.com/aarogya/images/malaria-activity.jpg

The *Anopheles* mosquito can be recognized by its upturning tail. Anopheles mosquitoes lay their eggs in water, which hatch into larvae, eventually emerges into adult mosquitoes. The female mosquitoes seek a blood meal to nurture their eggs. The eggs are very small but can be seen as small (2-5mm wide) black spots on the surface of water. The malaria mosquito chooses stagnant or slow-flowing water in which to lay her eggs. Each species of *Anopheles* mosquito has its own preferred aquatic habitat such as in rainwater pools and puddles, borrow pits, river bed pools, irrigation channels, seepages, rice fields, wells, pond margins, sluggish streams with sandy margins. After two or three days a mosquito larva will come out of each egg. The larva feeds on microscopic organisms and plants in the water, and grows until it becomes a pupa. The pupa remains in the water, but does not feed. After a few days, the adult mosquito will come out of the pupa and fly away. Then it lays eggs after taking blood meal and everything starts all over again. It takes 7-14 days for a mosquito to grow from an egg to an adult mosquito.

11.7. TRANSMISSION

Malaria is transmitted through the bite of an infected, female Anopheles mosquito between dusk and dawn and occasionally through blood transfusion. When a mosquito bites a person it sucks up blood. If the person has infected with malaria, some of the parasites in the blood will be sucked into the mosquito. The malaria parasites multiply and develop in the mosquito. After 10-14 days they are mature and ready to be passed on to someone else. If the mosquito now bites a healthy person, the malaria parasites enter the body of a healthy person. The parasites are transported in the bloodstream to the victim's liver where they multiply and then re-enter the bloodstream. The malaria parasites can multiply 10 times every 2 days, destroying red blood cells and infecting new cells throughout the body. The person infected by the mosquito bite will become ill with malaria, symptoms appears within 7-21 days after infection.

Transmission also depends on climatic conditions that may affect the number and survival of mosquitoes, such as rainfall patterns, temperature and humidity. In many places, transmission is seasonal, with the peak during and just after the rainy season. Malaria epidemics can occur when climate and other conditions suddenly favors transmission in areas where people have little or no immunity to malaria. People with low immunity also get infected when they move into areas with intense malaria transmission. Partial immunity is developed over years of exposure although it never provides complete protection but reduces the risk of severe infection.

11.8. SYMPTOMS

Typically, malaria causes fever, headache, vomiting and other flu-like symptoms.

The parasite infects and destroys red blood cells resulting in easy fatigue-ability due to anemia, fits/convulsions and loss of consciousness.

Parasites are carried by blood to the brain (cerebral malaria) and to other vital organs.

Malaria in pregnancy poses a substantial risk to the mother, the fetus and the newborn infant. Pregnant women are less capable of coping with and clearing malaria infections, adversely affecting the unborn fetus.

In severe cases following symptoms appears:

- Prostration (inability to sit), altered consciousness lethargy or coma
- Breathing difficulties
- Severe anaemia
- Generalized convulsions/fits
- Inability to drink/vomiting
- Dark and/or limited production of urine

11.9. DIAGNOSIS AND TREATMENT

The number of malaria cases worldwide seems to be increasing, due to increasing transmission risk in areas where malaria control has declined, the increasing prevalence of drug resistant strains of parasites, and in a relatively few cases, massive increases in international travel and migration. The need for effective and practical diagnostics for global malaria control is increasing, since effective diagnosis reduces both complications and mortality from malaria. Differentiation of clinical diagnoses from other tropical infections, based on patients' signs and symptoms or physicians' findings, may be difficult. Therefore, confirmatory diagnoses using laboratory technologies are urgently needed.

Early diagnosis and treatment of malaria reduces disease and prevents death cases. It also contributes to reducing malaria transmission. There are two ways malaria can be diagnosed: Microscopy and rapid diagnostic tests (RDTs).

Microscopy- a blood sample is taken from the patient and is looked at under the microscope. If parasites are visible within the blood smear they are diagnosed as having malaria. The key limitation is that this method of diagnosis can only be used in laboratories where there is electricity and trained medical staff.

RDTS- It is a quick test that use a drop of blood from the fingertip to identify if the patient has malaria. The tests are sensitive to antigens (proteins that are produced by the parasite) that bind with a dye to form a coloured strip (a bit like a pregnancy test) to indicate whether there are parasites in the blood. An RDT takes just 15 minutes and can be used in rural communities by trained community workers, making this a valuable and lifesaving diagnostic tool.

It is particularly important to make an early diagnosis of malaria in young children and in pregnant women. These two groups may rapidly become very ill and may die within a few days. Pregnancy reduces the immune status of individuals and hence makes them more susceptible to malaria infection. Malaria during pregnancy is quite difficult to treat as the parasites tend to hide themselves in the placenta, making diagnosis and treatment difficult.

Anti-malarial drugs kill the parasite but do not prevent the patient from being reinfected. Early and proper treatment of malaria with effective anti-malarial drugs can shorten the duration of the infection and prevent further complications which could belief threatening. WHO recommends Artemisinin combination therapies (ACTs) to treat *Plasmodium falciparum* infection. Drugs such as chloroquine and primaquine are recommended for *P.vivax* malaria.

Pregnant women

Malaria in a pregnant woman increases the risk of maternal death, miscarriage, stillbirth and low birth weight with associated risk of neonatal death. Pregnant women are particularly susceptible to mosquito bites and should therefore use protective measures, including insect repellents and insecticide-treated mosquito nets. Clindamycin and quinine are considered safe, including during the first trimester of pregnancy. Chloroquine can be safely used for treatment of *vivax* malaria during pregnancy.

Children

Chloroquine can be safely given to treat *P. malariae*, *P. ovale* or *P. vivax* infections in young children. Artemisinin-based combination therapy (ACT) as per national policy may be used as first-line treatment against *P.falciparum*.

11.10. PRECAUTIONS

Avoid being bitten by mosquitoes, especially between dusk and dawn.

Take antimalarial drugs (Chemoprophylaxis) when appropriate, at regular intervals to prevent acute malaria attacks.

Immediately seek diagnosis and treatment if a fever develops 1 week or more after entering an area where there is a malaria risk and up to 3 months (or, rarely, later) after departure from a risk area.

11.11. PREVENTIVE MEASURES

1. Early case Detection and Prompt Treatment (EDPT)

EDPT is the main strategy of malaria control - radical treatment is necessary for all the cases of malaria to prevent transmission of malaria.

Chloroquine is the main anti-malaria drug for uncomplicated malaria.

Drug Distribution Centers (DDCs) and Fever Treatment Depots (FTDs) have been established in the rural areas for providing easy access to anti-malarial drugs to the community.

Alternative drugs for chloroquine resistant malaria are recommended as per the drug policy of malaria.

2. Vector Control

(i) Chemical Control

Use of chemical larvicides like Abate, Temephos and baytex in potable water

Aerosol space spray during day time

Indoor residual spraying (IRS) with insecticides is a powerful way to rapidly reduce malaria transmission. DDT, Synthetic Pyrethroids, Malathion fogging in the indoor of houses is the most effective measure to kill the adult mosquito.

The use of mosquito repellents, protective clothing, bed nets, mosquito coils, screening of houses, etc also use as preventive measure against malaria.

(ii) Biological Control
- Use of larvivorous fish like Guppy in ornamental tanks, fountains and ponds etc.
- Use of biocides.
- (iii) Personal Prophylatic Measures that individuals/communities can take up
- Use of mosquito repellent creams, liquids, coils, mats etc.
- Screening of the houses with wire mesh
- Use of bednets treated with insecticide
- Wearing clothes that cover maximum surface area of the body

4. Community Participation
- Sensitizing and involving the community for detection of *Anopheles* breeding places and their elimination
- NGO schemes involving them in programme strategies

5. Environmental Management and Source Reduction Methods
- Source reduction i.e. filling of the breeding places
- Proper covering of stored water
- Channelization of breeding source

6. Vaccine

No definitive malarial vaccine has been approved the world over against malaria. A number of vaccines of potential value in controlling malaria are currently under development.

11.12. ANTI MALARIAL CAMPAIGN

Since, Malaria is a global problem to some extent but certain countries like India face a widespread infection of this disease. However with the assistance of World Health Organisation (WHO), the Ministry of Health of Government of India started a National Malaria Control Programme (NMCP) in year 1953. Vector control is an essential part for reducing malaria transmission and became less effective in recent years, due to many technical and administrative reasons, including poor or no adoption of alternative tools. Under NMCP programme effective measures were taken and malaria was almost controlled because DDT and other insecticides used were very much effective in eradicating the mosquitos. But in recent years the cases of malaria are frequently witnessed and the frequency is rapidly increasing again. It appears that the insects have developed resistance and imunity to DDT and other similar insecticides , they have also changed their behaviour. However various research laboratories in our country like Vector Control Research Centre at Pondicherry, National Institute of Communicable Diseases at Delhi and other are engaged in finding out the measures to check malaria infection and also different ways to eradicate the mosquitoes.

REFERENCES

- Brian M. Greenwood, David A. Fidock, Dennis E. Kyle, Stefan H.I. Kappe, Pedro L. Alonso, Frank H. Collins, Patrick E. Duffy. Malaria: progress, perils, and prospects for eradication. J. Clin. Invest. 2008;118:1266–1276. doi:10.1172/JCI33996 Full Text at http://www.jci.org/articles/view/33996/files/pdf
- Laurence Floren, Michael P. Washburn, J. Dale Raine et al. A proteomic view of the *Plasmodium falciparum* life cycle Nature October 2002;419:520-526. Full text at http://www.nature.com/nature/journal/v419/n6906/pdf/nature01107.pdf
- National Vector Borne Disease Control Programme (NVBDCP, 2017) :(http://nvbdcp.gov.inden-cd.html).
- WHO (2016) World Malaria report,(http://www.who.int/malaria/media/world-malaria-report-2016/en/).
- Centers for Disease Control and Prevention(CDC), https://www.cdc.gov/malaria/

12. AMOEBIASIS

T. K. Barik
T. Sarita Achari
Sasmita Panda

12.1. INTRODUCTION

Amoebiasis, also called amoebic dysentery or entamoebiasis or bloody flux, is a diarrheal illness caused by a parasite, *Entamoeba histolytica*. This parasite lives in the intestines. This protozoan parasite has a global distribution and an especially high prevalence in countries where poor socioeconomic and sanitary conditions predominate. In resource-rich nations infections may be seen in travelers and emigrants from endemic areas. Most infections are asymptomatic, but tissue invasion may result in amoebic colitis, life-threatening hepatic abscesses, and even hematogenous spread to distant organs. Importantly, disease can occur months to years after exposure and must remain in the differential diagnosis at risk populations. According to WHO, intestinal amoebiasis caused by *Entamoeba histolytica* is the third principal parasitic disease responsible for mortality in the world. This protozoal parasite infects approximately 180 million individuals throughout the world, among whom 40 to 110 thousand die from it each year.

12.2. CAUSATIVE AGENT

Entamoeba histolytica is an enteric protozoan parasite with worldwide distribution. It is the causative agent of amoebic dysentery (bloody diarrhea) and invasive extraintestinal amebiasis (such as liver abscess, peritonitis, pleuropulmonary abscess).

12.3. LIFE CYCLE OF ENTAMOEBA HISTOLYTICA

E. histolytica is a monogenetic parasite as its life cycle is completed in a single host i.e., man. Three distinct morphological forms exist in its life cycle. – Trophozoite, Pre-cystic stage and Cystic stage.

Trophozoite

It is the growing or feeding stage of the parasite. During this stage the parasite resides in the mucosa and sub-mucosa layers of the large intestine of man. Trophozoites are unicellular organisms, with a size ranging from 18 to 40 mm in diameter. During the optimal living condition the parasite exhibit slow gliding movement by forming pseudopodia, hence the body shape is not fixed because of constantly changing position. The cytoplasm inside the body of trophozoite is divisible into clear, transparent ectoplasm and inner granular endoplasm. The endoplasm contains nucleus, ingested red blood cells and tissue debris. A single spherical nucleus lies inside the endoplasm. The size of the nucleus ranges from 4 to 6 μm. Nucleus contains a central dot like karyosome and a delicate single layered nuclear membrane containing fine chromatin granules. The space between karyosome and the nuclear membrane is traversed by radially arranged fine threads of limn network. Trophozoite secretes a proteolytic ferment around itself. This ferment is of the nature of histolysin which brings about destruction and necrosis of the surrounding host tissues to be absorbed later by the parasite as food. Trophozoite reproduces by binary fission and increases their number. They are exclusively parasitic in nature, growing at the expense of living tissues and multiplying rapidly to maintain their presence in good number.

Pre- Cystic stage

It is an intermediate stage between the trophozoite and cystic forms. During this stage the parasite reduces in size (10-20 μm), becomes ovoid in shape and bears a single blunt pseudopodia. The endoplasm does not contain ingested R.B.C's and other tissue debris, indicating that during this stage the parasite stop feeding. A single nucleus remains present.

Cystic stage

Cyst formation occurs inside the lumen of the host's intestine. The precystic parasite moves into the gut lumen to be transformed into cystic form, a process called "encystations". During the process of encystations, the parasite becomes round and get surrounded by a double refractile wall, called the cyst wall. A cyst in the beginning is uninucleate body with size ranging from 7—15 um, in different races. The nucleus inside the cyst soon divides by binary fission to become a binucleate form and then to quadrinucleate form. In this way, a single nucleus by mitotic division forms four daughter nuclei, undergoes reduction in size and ultimately becomes two in diameter. Inside the cytoplasm of the cyst develops certain extra nuclear bodies like chromatid bars and glycogen mass. Chromatid bars or chromatoids are dark oblong bar like structures varying in size and number. In addition to chromatid bars the cyst also contains mass of glycogen in the form of brown vacuolar structure.

As the cyst transform from uninucleate to quadrinucleate stage, both chromatid bars and glycogen vacuole reduces in size and finally disappear. The whole process of encystation occurs within a few hours. The life of a matured cyst (quadrinucleate form) inside the lumen of the host's gut is only two days. The mature quadrinucleate cysts pass out of their host's body through faeces. Outside the body of the host, the cyst survives for ten days and their thermal death point is about 50°c.

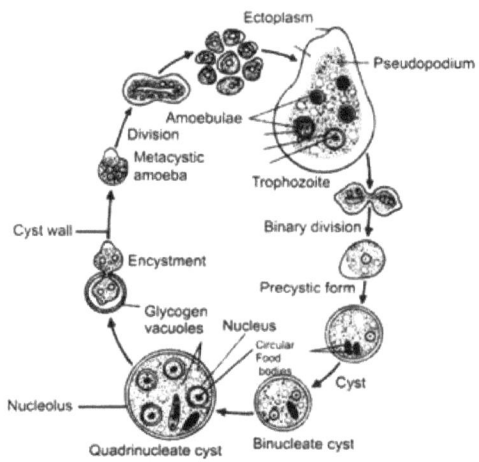

Figure 12.1

Life cycle of *E. histolytica*
Source: (http://www.apiindia.org/medicine_update_2013/chap01.pdf)

12.4. RESERVOIR AND SOURCE

Reservoir is human, an infected person suffering from the disease or an asymptomatic carrier discharging eggs in stools. Sources are water or food contaminated by excreta of patient or carries containing the cysts of *E. histolytica*.

12.5. TRANSMISSION

A matured quadrinucleate cyst of *Entamoeba histolytica* is the infective stage of the parasite. Transmission of *E. histolytica* from one person to another occurs due to ingestion of these cysts. Faecal contamination of edible substances and drinking water are the primary cause of infection. Following are the mode of transmission of this parasite-

12.5.1. Faecal-oral route

In majority of cases infection takes place through intake of contaminated uncooked vegetables and fruits. Insect vectors like flies, cockroaches and rodents act as agent to carry infective cysts to the food and drink. Sometimes drinking water supply contaminated with infected faces give rise to epidemics.

12.5.2. Oral-rectal contact

Sexual transmission by oral-rectal contact is also one of the modes of transmission, especially among male homosexuals.

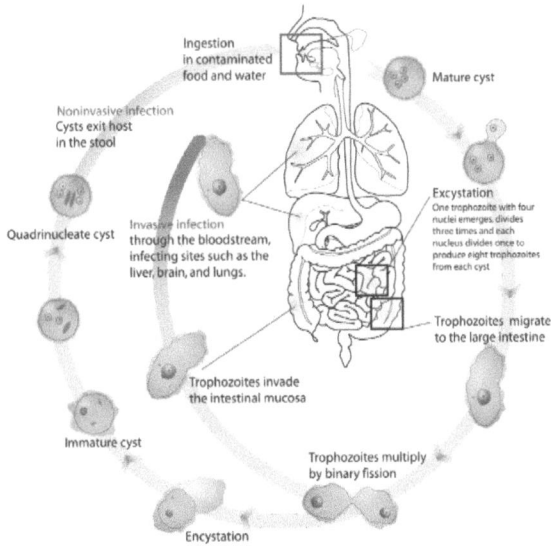

Figure 12.2

Source: (http://www.apiindia.org/medicine_update_2013/chap01.pdf)

12.6. PATHOGENESIS AND PATHOLOGY

Trophozoites are digested and destroyed by the gastric acid, hence, cannot cause infection even though they are ingested. The cyst divides into four initially, which divide again into eight daughter amoebae after an incubation period of 1–4 weeks, which may, however, be from few days to a year. These grow and mature into adult amoebae in about 7–10 days and stay as boarders in the large intestine, mainly the cecum and the sigmoid, feeding on intraluminal cellular debris and the bacteria. The infection is usually asymptomatic. Under unfavorable conditions and as the liquid stool becomes solid during its passage down the colon, the vegetative forms become cysts and are passed in the feces. Most individuals are asymptomatic cyst shredders. However, depending on the genetic and immune enzymatic profile and the parasite's ability to produce proteolytic enzymes, enabling resistance to complement-mediated lysis, the trophozoite becomes virulent and starts invading the intestinal mucosa. The trophozoites enhance the mucous secretion, alter its composition and deplete goblet cells of mucin, thereby making epithelial surfaces more vulnerable to invasion. Following depression of the mucous blanket, trophozoites attach to the cells of interglandular epithelium, and with the aid of proteolytic enzymes that degrade elastin, collagen and fibronectin, they invade the colonic epithelium by disruption of the extracellular matrix. Although there is evidence to suggest that *E. histolytica* can induce apoptosis of host cells, cell damage is primarily contact dependent. The first sign of colonic aggression may be manifested as nonspecific thickening of the mucosa or by pin-head size micronodules visible by sigmoidoscopy. Peptide-mediated lysosome enzymes released by the lysed polymorphonuclear leukocytes and monocytes contribute to the destruction of host tissue and extend the lesion. The microulcers coalesce extending the lesion further, undermining other areas of the colonic mucosa. With continued expansion, trophozoites invade the submucosa and spread out laterally, creating the classical flask shaped amoebic ulcers. Histopathology shows necrotic areas and vascular congestion. There is, however, little inflammation in contrast to the extension of the lesion. The amoebae may be found on the surface of the ulcers. Bacterial infection causes deficiency of blood supply leading to necrosis, haemorrhage, gangrene and subsequent perforation. Vascular invasion results in systemic spread. Once inside the portal vasculature, amoebae are resistant to complement mediated lysis. Right lobe of the liver is commonly involved, since blood draining the cecum, through a hypothetical laminar flow in portal vein, preferentially goes there. Left lobe involvement is fraught with the danger of pericardial involvement and sudden tamponade. Occasionally, small multiple abscesses occur mimicking bacterial abscesses. *E. histolytica* has the capacity to destroy almost all tissues of the human body, the intestinal mucosa, liver and to lesser extent the brain, skin, cartilage and even bone. The virulence factors, though not known with certainty, include adhesion molecules, proteases, hemolysins, contact-dependent cytolysis, apart from its phagocytic activity.

12.7. SYMPTOMS

When symptoms occur, they tend to appear one to four weeks after ingestion of the cysts. According to the Centers for Disease Control and Prevention (CDC), only about 10 to 20 percent of people who have amoebiasis become ill from it. Symptoms at this stage tend to be mild and include loose stools and stomach cramping.

If the parasite invades the lining of the intestine, it can produce amebic dysentery. Amoebic dysentery is a more dangerous form of the disease with frequent watery and bloody stools and severe stomach cramping. If the parasite enters your blood stream, it can end up in your liver, heart, lungs, brain, or other organs, where it causes tissue destruction and abscesses. The liver is a frequent destination for the parasite. Symptoms of amoebic liver disease include fever and tenderness in the upper-right part of your torso.

12.8. DIAGNOSIS

In developing countries, amoebiasis is usually diagnosed by the detection of motile trophozoites or cysts on a saline wet mount from a stool specimen. This is associated with a low sensitivity and high false-positive rate due to the finding of nonpathogenic parasites in the stool. Several alternative tests are used to diagnose intestinal amebiasis. The diagnosis is frequently made by the detection of *E. histolytica*-specific antigens or DNA in the stool and by the presence of antiamoebic antibodies in the serum. The antibody titer may remain positive for years. Sigmoidoscopy may reveal the characteristic ulcers, especially in more severe disease.

Aspirates or biopsies can also be examined microscopically for trophozoites. Several antigen detection kits are currently available and protocols for extracting fecal DNA and carrying out PCR are available. Serology is especially useful for the diagnosis of extraintestinal amebiasis. Seventy to eighty percent of patients with acute invasive colitis or liver abscesses, and greater than 90% of the convalescence patients, exhibit serum antibodies against *E. histolytica*. Noninvasive imaging techniques (e.g., ultrasound, computerized tomography, magnetic resonance imaging) can be used to detect hepatic abscesses. The detection of a space-occupying lesion in the liver combined with positive serology provides a high level of sensitivity and specificity. It is also possible to aspirate hepatic abscesses. However, this is rarely done and only indicated in selected cases (e.g., serology and imaging not available, therapeutic purposes). The aspirate is usually a thick reddish brown liquid that rarely contains trophozoites. Trophozoites are most likely to be found at the abscess wall and not in the necrotic debris at the abscess center.

12.9. TREATMENT

Colonization with *E. histolytica* should be treated with a luminal agent alone. Oral drugs that are effective against luminal infection include diloxanide furoate, paromomycin, and iodoquinol. The recommended duration of treatment with paromomycin is 7 days, with diloxanide furoate is 10 days, and with iodoquinol is 20 days. In a case in which luminal agents cannot be used, it seems a reasonable (if unproven) approach to treat luminal infection with metronidazole and test for cure with the stool antigen detection test. Invasive amebiasis (e.g., colitis, liver abscess) should be treated with metronidazole for 10 days. Although metronidazole has some unpleasant side effects, such as headache, nausea, metallic taste, and a disulfiram-like reaction to alcohol, reaction is rarely severe. Uncommon neurological side effects, such as vertigo or encephalitis, or neutropenia may necessitate discontinuation of treatment. Therapy with metronidazole should be followed with a luminal agent, since patients are otherwise at risk of relapsing from residual infection in the intestine. The majority of patients with amebic liver abscess defervesce after 3–4 days of treatment with metronidazole. Chloroquine and/or percutaneous drainage of the liver abscess are options in addition to metronidazole treatment for the rare patient who does not respond to metronidazole alone.

12.10. PREVENTION

Amoebiasis can be prevented and controlled both by non-specific and specific measures.

Non-specific measures are concerned with-

Improved water supply– The cysts are not killed by chlorine in amount used for water disinfection. Water filtration and boiling are more effective than chemical treatment of water against amoebiasis.

Sanitation–Safe disposal of human excreta coupled with the sanitary practice of washing hands after defecation and always before handling and consuming food.

Food safety– Uncooked fruits and vegetables should be washed thoroughly with safe water, peel fruits, and boil vegetables prior to eating. Measures should also include the protection of food and drink from flies and cockroaches and the control of these insects. Carriers, who pass cysts and are involved in handling food, whether at home, at street stalls, or in catering establishments, should be actively detected and treated since they are major transmitters of amoebiasis.

Health education of the public as well as health personnel at all levels about sanitation and food hygiene-Elementary hygienic practices should be propagated and constantly reinforced in schools, health care units, and the home through periodic campaigns using the mass media.

General social and economic development-The implementation of individual and community preventive measures (e.g., washing of hands, proper excreta disposal) should be an essential part of these activities.

Specific measures that should be undertaken when possible are-

Community surveys to monitor the local epidemiological situation with regard to amoebiasis;

improvement of case management, i.e., rapid diagnosis and adequate treatment of patients with invasive amoebiasis at all levels of the health services, including the community and health centre levels; Surveillance and control of situations that may encourage the further spread of amoebiasis, e.g., refugee camps, contaminated public water sources.

REFERENCES

- Beeching, Nick; Gill, Geoff (2014-04-17). "19". Lecture Notes: Tropical Medicine. John Wiley & Sons. pp. 177–182.
- Farrar, Jeremy; Hotez, Peter; Junghanss, Thomas; Kang, Gagandeep; Lalloo, David; White, Nicholas J. (2013-10-26). Manson's Tropical Diseases. Elsevier Health Sciences. pp. 664–671.
- Haque, Rashidul; Huston, Christopher D.; Hughes, Molly; Houpt, Eric; Petri, William A. (2003-04-17). "Amebiasis". NEJM. 348 (16): 1565–1573.
- Mondal D, Petri Jr WA, Sack RB, et al. (2006). "Entamoeba histolytica-associated diarrheal illness is negatively associated with the growth of preschool shildren: evidence from a prospective study". Trans R Soc Trop Med H. 100 (11): 1032–38.
- Recavarren-Arce S, Velarde C, Gotuzzo E, Cabrera J (March 1999). "Amoeba angeitic lesions of the central nervous system in Balamuthia mandrilaris amoebiasis". Hum. Pathol. 30 (3): 269–73.
- Visvesvara GS, Moura H, Schuster FL (June 2007). "Pathogenic and opportunistic free-living amoebae: Acanthamoeba spp., Balamuthia mandrillaris, Naegleria fowleri, and Sappinia diploidea". FEMS Immunol. Med. Microbiol. 50 (1): 1–26.
- KVGK Tilak (2013) .Amoebiasis , Infectious Diseases
- (http://www.apiindia.org/medicine_update_2013/chap01.pdf)

13. DIARRHOEA

T.K. Barik

Simani Mohanty

Sasmita Panda

13.1. INTRODUCTION

The World Health Organization defined diarrhoea as having three or more loose or liquid stools per day, or as having more stools than is normal for that person. Acute diarrhea is defined as the abrupt onset of 3 or more loose stools per day and lasts no longer than 14 days; chronic or persistent diarrhea is defined as an episode that lasts longer than 14 days. The most common cause of acute diarrhoea is an infection of the intestines, such as gastroenteritis or food poisoning. Viruses are responsible for most cases. The intestinal lining becomes irritated and inflamed, which hinders the absorption of water from food waste. In severe cases, the intestinal lining may even leak water. Diarrhoea is one of the most common reasons for people to seek medical advice; it can range from a mild, temporary condition, to one that can be life-threatening. Globally, there are about 2 billion cases of diarrhoea disease each year, and 1.9 million children under the age of 5, mostly in developing countries, die from diarrhoea. Diarrhoea should not be confused with the frequent passing of stools of normal consistency - this is not diarrhoea. Diarrhoea is characterized by abnormally loose or watery stools. Similarly, breastfed babies often pass loose, pasty stools, which is normal and not diarrhoea.

Diarrhoea disease may have a negative impact on both physical fitness and mental development. Acute diarrhoea can be life threatening to babies and young children. This is because their smaller bodies are more vulnerable to dehydration.

13.2. TYPES OF DIARRHOEA

13.2.1. Secretory

Secretory diarrhoea means that there is an increase in the active secretion, or there is an inhibition of absorption. There is little to no structural damage. The most common cause of this type of diarrhoea is a cholera toxin that stimulates the secretion of anions, especially chloride ions. Therefore, to maintain a charge balance in the gastrointestinal tract, sodium is carried with it, along with water. In this type of diarrhoea intestinal fluid secretion is isotonic with plasma even during fasting. It continues even when there is no oral food intake.

13.2.2. Osmotic

Osmotic diarrhoea occurs when too much water is drawn into the bowels. If a person drinks solutions with excessive sugar or excessive salt, these can draw water from the body into the bowel and cause osmotic diarrhoea. Osmotic diarrhoea can also be the result of mal-digestion (e.g., pancreatic disease or Coeliac disease), in which the nutrients are left in the lumen to pull in water. In healthy individuals, too much magnesium or vitamin C or undigested lactose can produce osmotic diarrhoea and distention of the bowel. A person who has lactose intolerance can have difficulty absorbing lactose after an extraordinarily high intake of dairy products. In persons who have fructose mal-absorption, excess fructose intake can also cause diarrhea. High-fructose foods that also have a high glucose content are more absorbable and less likely to cause diarrhoea. Sugar alcohols such as sorbitol (often found in sugar-free foods) are difficult for the body to absorb and, in large amounts, may lead to osmotic diarrhoea. In most of these cases, osmotic diarrhoea stops when the offending agent (e.g. milk, sorbitol) is stopped.

13.2.3. Exudative

Exudative diarrhoea occurs with the presence of blood and pus in the stool. This occurs with inflammatory bowel diseases, such as Crohn's disease or ulcerative colitis, and other severe infections such as *E. coli* or other forms of food poisoning.

13.2.4. Inflammatory

Inflammatory diarrhoea occurs when there is damage to the mucosal lining or brush border, which leads to a passive loss of protein-rich fluids and a decreased ability to absorb these lost fluids. Features of all three of the other types of diarrhoea can be found in this type of diarrhoea. It can be caused by bacterial infections, viral infections, parasitic infections, or autoimmune problems such as inflammatory bowel diseases.

13.2.5. Dysentery

If there is blood visible in the stools, it is also known as dysentery. The blood is a trace of an invasion of bowel tissue. Dysentery is a symptom of, among others, Shigella, *Entamoeba histolytica*, and *Salmonella*.

It can cause symptoms electrolyte imbalances, renal impairment, dehydration, and defective immune system responses. When oral drugs are administered, the efficiency of the drug is to produce a therapeutic effect and the lack of this effect may be due to the medication travelling too quickly through the digestive system, limiting the time that it can be absorbed. Clinicians try to treat the diarrhoea by reducing the dosage of medication, changing the dosing schedule, discontinuation of the drug, and rehydration. The interventions to control the diarrhoea are not often effective. Diarrhoea can have a profound effect on the quality of life because faecal incontinence is one of the leading factors for placing older adults in long term care facilities (nursing homes).

13.3. DIAGNOSIS

The doctor will start by asking questions about the problem, including current medications, past medical history, and other medical conditions. They will also ask:
- When the problem started.
- Stool frequency, type (for example, watery, mucus-filled, pussy) and volume.
- Whether blood is present in the stool.
- Whether there has been vomiting.

Doctors will also be concerned about whether there is dehydration. Severe dehydration can be fatal if treatment with rehydration therapy is not given urgently. Because most cases of diarrhoea resolve themselves, and because the diagnosis can be made clinically, tests are not usually required. But in more severe cases, for example, doctors may order further testing. Acute cases, particularly if the patient is very young or old, may require a stool sample to be tested. Other factors might also need to be investigated; for instance, if the patient:
- Has signs of fever or dehydration.
- Has stools with blood or pus.
- Has severe pain.
- Has low blood pressure.
- Has recently traveled to places outside Western Europe, North America, Australia, and New Zealand.
- Has recently received antibiotics or been in hospital.
- Has diarrhoea persisting for more than 1 week.
- Chronic cases of diarrhoea will be tested according to the suspected underlying cause, and may include the following investigations:
- Full blood count - for anemia or a raised platelet count which suggests inflammation.
- Liver function tests, including albumin level.
- Tests for mal-absorption - calcium, vitamin B12, folate, iron status (ferritin), thyroid function tests.
- ESR (erythrocyte sedimentation rate) and CRP (C-reactive protein) - with raised levels possibly pointing to inflammatory bowel disease (IBD).
- Celiac disease testing - for antibodies.

13.4. CAUSES:

Most cases of diarrhoea are caused by an infection in the gastrointestinal tract. The microbes responsible for this infection include:

Bacteria

Viruses

Parasitic organisms

The most commonly identified causes of acute diarrhea in the United States are the bacteria *Salmonella, Campylobacter, Shigella*, and Shiga-toxin-producing *Escherichia coli*.

Some cases of chronic diarrhoea are called "functional" because a clear cause cannot be found. In the developed world, irritable bowel syndrome (IBS) is the most common cause of functional diarrhoea.

IBS is a complex of symptoms. There is cramping abdominal pain and altered bowel habits - diarrhoea, constipation, or both.

Inflammatory bowel disease (IBD) is another cause of chronic diarrhoea. It is a term used to describe either ulcerative colitis or Crohn's disease. There is often blood in the stool in both conditions.

Other major causes of chronic diarrhoea include:
1. Microscopic colitis - usually affects older adults. The persistent diarrhoea is often during the night.
2. Malabsorptive and maldigestive diarrhoea - the first is caused by impaired nutrient absorption, the second by impaired digestive function. Celiac disease is one example.
3. Chronic infections - a history of travel or antibiotic use can be clues to chronic diarrhoea; various bacteria and parasites can be the cause.
4. Drug-induced diarrhoea - the obvious cause is laxatives, but a list of other drugs can also lead to diarrhoea including antibiotics.
5. Endocrine causes - sometimes hormones are the cause, for example, in conditions including Addison disease and carcinoid tumors.
6. Cancer causes - neoplastic diarrhoea is associated with a number of gut cancers.

13.5. PREVENTION

In developing countries, prevention of diarrhoea may be more challenging due to dirty water and poor sanitation. The following practical measures help to prevent the condition:

Clean/safe drinking water.

Good sanitation (toilets and sewerage).

Hand washing with soap - after defaecation, after cleaning a child who has defaecated, after disposing of a child's stool, before preparing food, and before eating.

For mothers with young babies, breast feeding for the first 6 months of life.

Good hygiene practices - both personal hygiene and in the kitchen.

Education on the spread of infection.

There is evidence that interventions from public health bodies to simply promote hand washing can cut diarrhoea rates by about one-third.

13.6. TREATMENT

1. Mild cases of acute diarrhoea may resolve without treatment. Persistent or chronic diarrhoea will be diagnosed and treated in addition to the symptoms of diarrhoea. For all cases of diarrhoea, the first important step in treatment is to rehydrate:
2. Fluids can be replaced by simply drinking more fluids, or they can be received intravenously in severe cases. Children and older people are more vulnerable to dehydration.
3. Oral rehydration solution/salts (ORS) - this is water that contains salt and glucose. It is absorbed by the small intestine to replace the water and electrolytes lost in the stool. In developing countries, ORS costs just a few cents; the World Health Organization (WHO) says ORS can safely and effectively treat over 90 percent of non-severe diarrhoea cases).
4. Oral rehydration products are available commercially - for example Oralyte and Rehydralyte. Zinc supplementation may reduce the severity and duration of diarrhoea in children.
5. OTC anti diarrhoeal medicines are also available:
6. Loperamide (Imodium, for example) is an anti-motility drug that reduces stool passage.
7. Bismuth subsalicylate (for example, Pepto-Bismol) reduces diarrhoea stool output in adults and children and may be a safer alternative to loperamide. This drug can also be used to prevent traveler's diarrhoea.
8. There is some concern that anti-diarrhoeal medications could prolong bacterial infection by reducing the removal of pathogens via stools.
9. Antibiotics are only used to treat diarrhoea caused by a bacterial infection. If the cause is a certain medication, switching to another drug might be possible.
10. Nutrition
11. Nutritionists from Stanford Health Care offer some nutritional tips for diarrhoea:
12. Sip on clear, still liquids such as fruit juice without added sugar, replacing lost water after each loose stool with at least one cup of liquid.
13. Do most of the drinking between, not during meals.
14. Consume high-potassium foods and liquids - examples include diluted fruit juices, potatoes (without the skin), bananas.
15. Use high-sodium foods and liquids - broths, soups, sports drinks, salted crackers.
16. Other advice from the nutritionists is to:
17. Eat foods high in soluble fiber to help thicken the stool - bananas, rice, oatmeal etc.
18. Limit certain foods that may make diarrhoea worse such as creamy, fried, and sugary foods.

19. Certain food and drink might make the diarrhoea worse:
20. Sugar-free gum, mints, sweet cherries, prunes
21. Caffeinated drinks and medication
22. Fructose in high amounts, from fruit juices, grapes, honey, dates, nuts, figs, soft drinks, and prunes.
23. Lactose in dairy products.
24. Magnesium.
25. Olestra (Olean) - a fat substitute.

REFERENCES

- Alam NH, Ashraf H (2003). "Treatment of infectious diarrhea in children". Paediatr Drugs. 5 (3): 151–65.
- DuPont HL (Apr 17, 2014). "Acute infectious diarrhea in immunocompetent adults.". The New England Journal of Medicine. 370 (16): 1532–40.
- Grantham-McGregor SM, Walker SP, Chang S (February 2000). "Nutritional deficiencies and later behavioural development". The Proceedings of the Nutrition Society. 59 (1): 47–54.
- Greenberg HB, Estes MK (May 2009). "Rotaviruses: from pathogenesis to vaccination". Gastroenterology. 136 (6): 1939–51.
- Guerrant RL, Schorling JB, McAuliffe JF, de Souza MA (July 1992). "Diarrhea as a cause and an effect of malnutrition: diarrhea prevents catch-up growth and malnutrition increases diarrhea frequency and duration". The American journal of tropical medicine and hygiene. 47 (1 Pt 2): 28–35.
- Longstreth GF, Thompson WG, Chey WD, Houghton LA, Mearin F, Spiller RC (2006). "Functional bowel disorders". Gastroenterology. 130 (5): 1480–91.
- Mitchell DK (November 2002). "Astrovirus gastroenteritis". The Pediatric Infectious Disease Journal. 21 (11): 1067–9.
- Moon, Changsuk; Zhang, Weiqiang; Sundaram, Nambirajan; Yarlagadda, Sunitha; Reddy, Vadde Sudhakar; Arora, Kavisha; Helmrath, Michael A.; Naren, Anjaparavanda P. (2015). "Drug-induced secretory diarrhea: A role for CFTR". Pharmacological Research. 102: 107–112.
- Navaneethan U, Giannella RA (November 2008). "Mechanisms of infectious diarrhea". Nature Clinical Practice Gastroenterology & Hepatology. 5 (11): 637–47.
- Patel MM, Hall AJ, Vinjé J, Parashar UD (January 2009). "Noroviruses: a comprehensive review". Journal of Clinical Virology. 44 (1): 1–8.

- Rossignol JF, Lopez-Chegne N, Julcamoro LM, Carrion ME, Bardin MC (2012). "Nitazoxanide for the empiric treatment of pediatric infectious diarrhea". Trans. R. Soc. Trop. Med. Hyg. 106 (3): 167–73.
- Rupnik M, Wilcox MH, Gerding DN (July 2009). "*Clostridium difficile* infection: new developments in epidemiology and pathogenesis". Nature Reviews Microbiology. 7 (7): 526–36.
- Uhnoo I, Svensson L, Wadell G (September 1990). "Enteric adenoviruses". Baillière's Clinical Gastroenterology. 4 (3): 627–42.
- Viswanathan VK, Hodges K, Hecht G (February 2009). "Enteric infection meets intestinal function: how bacterial pathogens cause diarrhoea". Nature Reviews Microbiology. 7 (2): 110–9.
- Wedlake L, A'Hern R, Russell D, Thomas K, Walters JR, Andreyev HJ (2009). "Systematic review: the prevalence of idiopathic bile acid malabsorption as diagnosed by SeHCAT scanning in patients with diarrhoea-predominant irritable bowel syndrome". Alimentary pharmacology & therapeutics. 30 (7): 707–17.

14. FILARIASIS

Chinmayee Panda
T. K. Barik

14.1. INTRODUCTION

Filariasis is caused by a round, coiled and thread-like parasitic worms belonging to the family *filaridea* transmitted by blood-feeding arthropods, mainly black flies and mosquitoes. Most cases of filariasis are by *Wuchereria bancrofti* parasite; the *Culex quinquefasciatus* and some species of Aedes and *Anopheles* mosquito transmit the disease. Another parasite called *Brugia malayi* that causes filaria is transmitted by the vector *Mansonia* and *Anopheles* mosquito. Therefore, mosquito is known as the vector host or intermediate host and Man is the natural host.

Lymphatic filariasis is the world's second leading cause of long-term disability. Although filariasis does not kill, it causes disability and imposes severe social and economic burden .The current estimate reveals that 120 million people in 83 countries of the world are infected with lymphatic filarial parasites, and more than 1.1 billion people are at risk of acquiring infection.

14.2. ORIGIN AND HISTORY

The disease was recorded in India as early as 6th century B.C. by Susruta, in his book 'Susruta Samhita' and in 7th century A.D., Madhavakara described sign and symptoms of the disease in histreatise 'Madhava Nidhana', which hold good even today. In 1709, Clarke called elephantoid legs in Cochin as 'Malabar legs'. The discovery of microfilariae (mf) in the peripheral blood was made first by Lewis in 1872 in Kolkata City of India.

14.3. CAUSATIVE AGENT

Nematode parasites causing *Lymphatic Filariasis* in human are *Wuchereria bancrofti, Brugia malayi* and *Brugia timori*. Of these, only *Wuchereria bancrofti* and *Brugia malayi* are found in India. *Wuchereria bancrofti*, transmitted by the ubiquitous vector, *Culex quinquefasciatus* has been the predominant infection. The vector species breeds preferably in dirty and polluted water. *Brugia malayi* infection restricted to some rural areas.

14.4. MORPHOLOGY OF THE PARASITE

The adult worms are slender and thread like in appearance. The body color is creamy-white. Anterior end of the body is swollen slightly and is distinguished as the head end. The mouth is in the form of a simple pore without lips and is situated at the anterior end. The females measure about 100mm.in length and about 0.24 to 0.3mm.in diameter. The males are comparatively smaller in size than the females and they measure about 40mm in length and about 0.1nm in diameter. There is a pair of copulatory spicules found in males out of the two copulatory spicules, one is longer than the other.

14.5. LIFE CYCLE OF THE PARASITE

The adult parasite worms, male and female, live in the lymph vessels and lymph nodes. The adult worms survive for about 5-8 years and sometimes for as long as 15 years. After mating, the female worm parturates millions of microfilariae which finally migrate to blood circulation. The sheathed microfilariae begin to appear in the blood circulation in six months to one year after infection (prepatent period). The microfilariae remain in the arterioles of the lungs during the day and emerge into the peripheral circulation at night (nocturnally periodic).

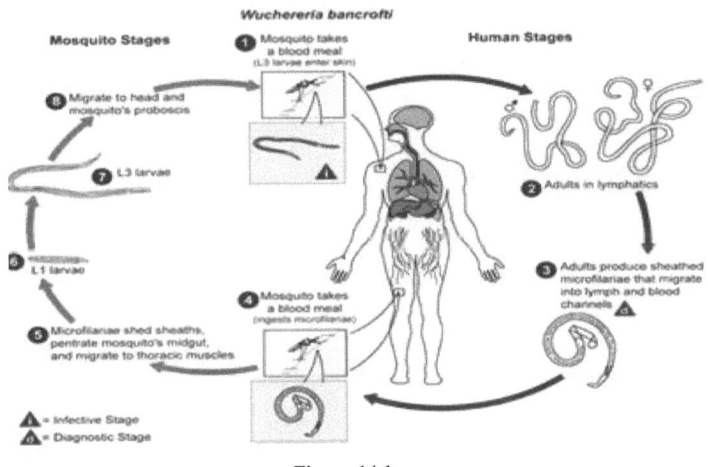

Figure 14.1

Source: http://www.medicalook.com/diseases_images/Filariasis.gif

The sexual cycle of the parasite takes place in the human host, where the adult worms ultimately die. The life cycle of the parasite is cyclo developmental in the vector where the parasites do not multiply. Microfilariae, (when picked up by the mosquito during blood meal) undergo development in mosquitoes (intermediate hosts) to form infective larvae which usually takes about 10 to 14 days. The ingested microfilariae first shed their sheaths, penetrate the stomach wall, migrate to the muscles of the thorax and develop there without multiplication. The slender and tiny microfilariae transform into immobile and inactive sausage stage L1 larva, which has a cuticle that forms a conspicuous slender tail with specific identifying characters. The larvae grow rapidly in length and breadth after their first moult to become L2 or pre-infective larva, which is recognized by the presence of one or two papillae at its caudal end and by its short tail. This L2 stage moults to become L3 which is infective. It is slender and thread like, measuring about 1500-2000 microns in length. It is highly motile which is a unique phenomenon used for identification. When the infective mosquitoes (harbouring L3 larvae) bite, some or all of the infective larvae escape from the proboscis and actively enter the human host through the wound made by the mosquito bite or penetrate the skin on their own and migrate into lymphatic system. In the lymphatic system of the infected persons, the infective larvae develop into adult male and female worms.

Several recent studies have demonstrated presence of Wolbachia, bacterial endosymbionts in the adult filarial worms and microfilariae of both W. bancrofti and B. malayi. This bacterium is necessary for the development, viability and fertility of the adult parasites. Drug interventions directed against Wolbachia cause deleterious effect on the survival of the adult worms.

14.6. DISTRIBUTION

According to the World Health Organization, few countries like India, Indonesia, Nigeria and Bangladesh alone contribute about 70% of the infection worldwide.

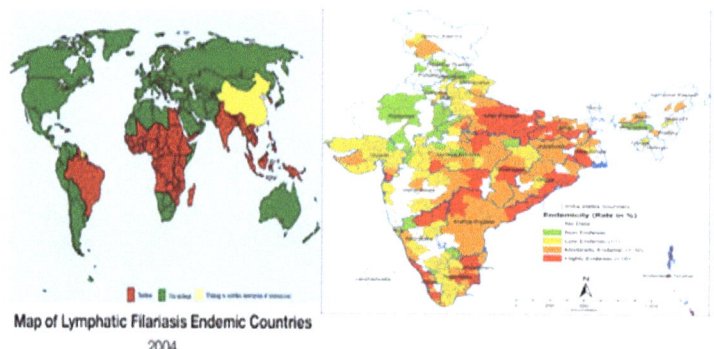

Map of Lymphatic Filariasis Endemic Countries
2004

Figure 14.2

Source:http://www.mosquitocontrol.org/~mosquito/sites/default/files/pictures/map_filariasis_world.jpg

14.7. CLASSIFICATION OF DISEASE

Depending on the area upon which the worms affect filaria is classified as

- Lymphatic filarial (*Elephantiasis*) affects the lymphatic system, including the lymph nodes.
- Subcutaneous filarial affects the subcutaneious layer of the skin.
- Serous cavity filarial affects the serous cavity of the abdomen.

14.8. FILARIA VECTORS

Culex quinquefasciatus mosquito is the vector of *W.bancrofti* in the mainland. *C. quinquefasciatus* breeds in association with human habitations and is the domestic pest mosquitoes, preferring polluted waters, such as sewage and sullage water collections including chess pools, chess pits, drains and septic tanks. In the absence of such type of water collections, they can breed in comparatively clean water collections also.

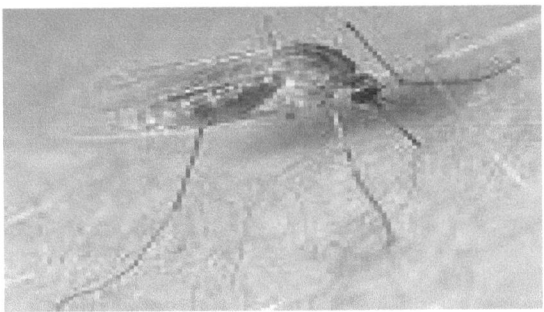

Figure 14.3

Culex quinquefasciatus

Source:at:http://www.mosquitocontrol.org

The eggs are laid in rafts containing 150-40 eggs each depending on quality and quantity of blood meal taken. At the optimum temperature of 25°C to 30°C, the eggs hatch within 24 to 48 hours. The youngest stage is the first instar larva which moults to subsequent instars each within 24-48 hours at optimum temperature. There are four instars in the larval stages, and all the instars are voracious eaters, taking anything and everything of microscopic size into the buccal cavity by instant vibration of its feeding brushes. They are mainly bottom feeders but may feed from the surface also.

The IV instar at the end of its stage gives rise to a comma shaped pupa, which lasts up to 24-48 hours at optimal condition. Pupae do not feed but are very active, respiring through its pair of breathing trumpets. The pupa emerges into an adult mosquito.Through a longitudinal slit formed between the two trumpets. The entire cycle from egg to emergence of adult is completed in 10-14 days.

14.9. CAUSE AND MODE OF INFECTION

There are mainly 8 different types of thread-like nematodes that cause filariasis.
- Lymphatic filaria is caused by Wuchereria bancrofti, Brugia malayi & Brugiatimari.
- Subcutaneous filaria is caused by *Loa loa* (the eye worm), *Mansonella streptocerca* & *Onchocerca volvulus*.
- Serous cavity filaria by the *woums mansonella parstans & mansonella ozzardi*.
- These parasites after getting deposited on skin penetrate on their own or through the opening created by mosquito bites to reach the lymphatic system.

14.10. SYMPTOMS

1. The disease causes no symptoms in the initial stage. Therefore, most people initially are not aware that they have filariasis. The symptoms are generally caused by the adult worms and not by the larvae(microfilariae).the lymph glands and lymph channels are enlarged in some cases with repeated infection, thearms, legs and certain parts of the male genital system such as scrotum spermatic cords, epididymis and testes ,also become enlarged.
2. These symptoms are due to the obstruction of lymph vessels, causing some degree of inflammation. This in turn causes increased amount of protein to enter the area, which stimulate the excessive growth of connective tissue.
3. Symptoms of lymphatic filariasis- Edema with thickening of the skin and underlying tissues. It usually affects the lower extremities. However, the arms, vulva, breast and scrotum (causing hydrocele formation) can also be affected. The edema in the extremities, breast or genital area can result in the part becoming several times its normal size and is due to blockage of the vessels of the lymphatic system.
4. Symptoms of subcutaneous filariasis- Include skin erashes, hyper or hypo pigmented mascules, riverblindness (caused by *onchocerca volvulus*)
5. Symptoms of serous filariasis- Abdominal pair, skin rashes.
6. Brugian filariasis- Lymphadenitis (swollen and painful lymph node) occurs episodically, most commonly affecting one inguinal lymph node at a time. The infection lasts for several days and usually heals spontaneously. The frequency of episodes may vary from 1-2 attacks per year to several attacks per month. Sometimes lymphadenitis is followed by a characteristic retrograde lymphangitis. The infection may spread to the surrounding tissues, and occasionally involves the whole thigh or entire limb. The infected lymph node may become an abscess, ulcerate, and heal with fibrotic scarring. The acute clinical course with its complications may last from several weeks to 3 months. Characteristically, elephantiasis involves the leg below the knee but occasionally it affects the arm below the elbow. Genital lesions or chyluria (milky color urine) do not occur in *brugian filariasis*.

14.11. BANCROFTIAN FILARIASIS

The lymphatic vessels of the male genitalia are most commonly affected in bancroftian filariasis, producing episodic funiculitis (inflammation of the spermatic cord), epididymitis and orchitis. Adenolymphangitis of the extremities is less common. Hydrocele is the most common sign of chronic bancroftian filariasis, followed by lymphoedema, elephantiasis and chyluria.

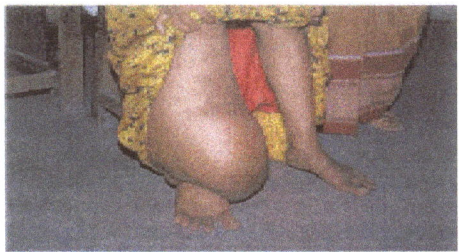

Figure 14.4

Source: http://www.nvbdcp.gov.in/images/filariasisleg.gif

The swelling involves the whole leg, the whole arm, the scrotum, the vulva or the breast. The fluid of hydrocele and chyluric patients may contain microfilariae, even when they are absent from the blood. Chyluria occurs intermittently and is more pronounced after a heavy meal. It is often symptomless, but some patients complain of fatigue and weight loss, resulting from loss of fat and protein.

14.12. LYMPHATIC FILARIASIS (LF)

Lymphatic Filariasis (LF), commonly known as elephantiasis is a disfiguring and disabling disease, usually acquired in childhood. In the early stages, there are either no symptoms or non-specific symptoms. Although there are no outward symptoms, the lymphatic system is damaged. This stage can last for several years. Infected persons sustain the transmission of the disease. The long term physical consequences are painful swollen limbs (lymphoedema or elephantiasis). Hydrocele in males is also common in endemic areas. Due to damaged lymphatic system, patients with lymphoedema have frequent attacks of infection causing high fever and severe pain. Patients may be bed-ridden for several days and normal routine activities become difficult. Such attacks not only cause acute physical suffering but also directly impede the earning capacity of the individual.

14.13. DIAGNOSIS

Commonly used method includes direct demonstration of microfile in blood or skin specimens. The blood sample is usually collected at night as microfilariae have nocturnal periodicity. Thick and thin blood films should be stained with Giemsa or haematoxylin and the microfilariae differential according to the pattern of their sheaths, nuclei distributions and size. The edges of the film should be examined carefully as microfilariae tend to be concentrated at the periphery and are easily missed if the microscopist goes straight onto high power in the center of film.

Detection of circulating antigen by enzyme linked immunosorbent assay (ELISA) or ICT has replaced Microscopy for the diagnosis of bancroftian, but not brugian, filariasis.

An antigen immunochromatography card test is available for the detection of W. bancrofti, which does not react with other filariae and is highly sensitive (100%) and specific (92%).

Filarial DNA can be detected by PCR, and ultrasound can help to identify adult worms within the lymphatic system. Serological tests are not very helpful in the diagnosis as most individuals from endemic areas have antibodies to crude filarial antigens and there is cross reactivity with other filariae and nematodes.

14.14. TREATMENT

Several antimony and arsenic compounds have been proved to be useful to some extent. Neostibosan (an antimony compound) destroys the microfilariae quickly, but has a slower effect on the adult worms. Arsenamide (an arsenic compound) kills the adults quickly, but has a slower effect on the microfilariae. The arsenamide is to be applied intravenously.

Diethylcarbamazine (DEC) is the recommended medicine for treating filaris. It kills microfilaria and does not have any effect on the adult worms.Thus,it only helps to control transmission of infection from one to other person.

Ivermection or albendozole may be useful in some patients.

Administer annual single dose of anti-filarial drugs (DEC or DEC+Albendazole). This annual dose is to be repeated every year for a period of 5 years or more aiming at minimum 85 % actual drug Compliance.

	STREAMLINED DOSE			
	DEC		ALBENDAZOLE	
AGE IN YEAR	DOSE(mg)	NO.OF100mg TABLET	DOSE(mg)	NO.OF100mg TABLET
LESS THAN 2 YEAR	0	0	0	0
2-5YEAR	100	1	400	1
6-14 YEAR	200	2	400	1
15 YEAR AND ABOVE	300	3	400	1

Table 14.1

Source: NVBDCP (2009)

14.15. INDIVIDUAL CHEMOTHERAPY

Diethylcarbamazine remains the treatment of choice. Adequate dosage will kill adult worms. Even a small single dose will clear blood microfilariae temporarily. Sensitivity reactions to filarial antigen, both local and systemic, are common in infected people and simulate some of the acute manifestation of the infection; they necessitate care and supervision in the initial stages, especially in Brugia infections.

Treatment should be started at 1 mg/kg on the first day, increasing over 3 or more days to 6 mg/kg in divided doses; this dose then being continued for 21 days. Coinfection with Loa loa and Onchocerca volvulas must be excluded before diethylcarbamazine is given to avoid dangerous reactions.

Indication for curative treatment is acute manifestation with or without microfilariaemia, and chronic disease in patients who are either microfilaria positive or positive for filarial antigen identified serologically. Treatment often reduces the size of Hydroceles but has little effect on chronic lymph edema.

14.16. SURGICAL AND SUPPORTIVE TREATMENT

Acute manifestation of filariasis can mimic strangulated hernia and testicular torsion. Surgical treatment of filarial Hydrocele is the same as that for non-filarial disease.

Scrotal lymphodema can be treated surgically, usually with preservation of the testes. Lymphosaphenous anastomosis is being used for leg elephantiasis.

Bacterial infection is common in those with lymphodema, especially when the skin is fissured, breached in an inter-digital cleft, or when there is minor injury, ulcer, or insect bite. Early use of antibiotics and resting of the affected limb lessens the risk of increasing lymphodema; supportive bandaging applied each morning or wearing elastic stocking reduces chronic oedema.

14.17. PREVENTION

These campaigns are targeted at the local vector. Larval Aedes breeding sites such a discarded tins, tyres, or coconut shells can be removed. Culex numbers can be reduced by improved sanitation, larvicides, and polystyrene beads applied to the water surface of latrines and cesspits. Bed nets and repellants are universally applicable. Where Anopheles is the vector, malaria control can interrupt filariasis transmission as in Samoa, Vanuatu, and part of southern China.

14.18. PERSONAL PRACTICE

Good hygiene of the affected part prevents the worsening of the lymphedema and secondary bacterial skin infection. The affected limb should be kept elevated and regular exercises should be done to improve the lymph flow.

Filariasis is not a life threating infection but it can cause lasting damage to the lymphatic system. While medicine like DEC are available to teat filariasis but the gross swelling of the leg makes a person look noticeable and ugly. Hence, it is better to protect from the bites of filarial spreading mosquito by using mosquito repellent creams, mats,coils,aerosols and prevent breeding of mosquito with better practice of hygiene and sanitation.

REFERENCES

- Guidelines on Filariasis Control in India and its Elimination - NVBDCP
- nvbdcp.gov.in/doc/guidelines-filariasis-elimination-india.pdf
- World Health Organization apps. who.int/iris/ bitstream/ 10665/97377/1/ 9789241505444_eng.pdf
- World Health Organization: *Lymphatic filariasis* apps.who.int/iris/bitstream /10665/ 87989/ 1/9789241505642_eng.pdf
- CG Health Portalcghealth.nic.in/ehealth/2013/QA_website/ 5Standard Treatment Guidelines/ Filaria.pdf
- National Vector Borne Disease Control Programme (NVBDCP) nvbdcp.gov.in/ filariasis-new.html

15. CANCER

G. K. Panigrahi

15.1. INTRODUCTION

Cell is the structural and functional unit of living organism. Cell is the basic unit of life which exhibit an advanced cellular organization in case of eukaryotes by containing diverse cell organelles. A multicellular organism grows well when all its molecular events regulating the cell growth and division are tightly regulated and thus the cellular environment thrives in a normal fashion. In rare cases, a normal cell suddenly behaves abnormally as if it became a rebel and shows a reluctance towards the fundamental rules that govern cell growth and reproduction. Then they start dividing hastily, invading other tissues, grabbing resources, and even in some cases leading into killing the body in which it lives. To dissect the causes that leads the cell to unfollow the basic principles can be understood only, when we can unsolved the mysteries related to the normal functions of cell growth and reproduction. Cutting edge research in basic and applied biosciences field has provided a detailed and specific information about the molecules and processes that plays a pivotal role in the cell cycle involving cell division, cell growth, cell differentiation that specifies the cells with their distinct roles. This detailed understanding provides an insight into various mechanisms that triggers cancer. The tight regulation of cell cycle plays a very important role in not allowing the normal cells to become cancerous thus loss of control of the cell cycle is one of the crucial steps in development of cancer. Even though, until now a number of different type of cancer diseases have been discovered, but they all share an important characteristic in common: they are abnormal cells where the normal cell division process is disturbed. Cancer is a consequence of switching of normal cells to abnormal cells which are triggered by various factors like inherited mutations, environmental factors such as UV light, X-rays, chemicals, tobacco products, and viruses. Studies suggests that most cancers are an outcome of various synchronized events and thus evolving the cell through a series of premalignant stages into an invasive cancer. The visible symptoms of cancer may appear years after initial event and the development of cancer. A high throughput sophisticated and reliable molecular biological techniques may help in the diagnosis of potential cancers in the early stages, well before the formation of any visible phenotypic symptoms.

15.2. WHAT IS CANCER?

Cancer is an outcome of a series of molecular events which are aimed to alter the physiology of normal cells by disabling the basic processes that prevent cell overgrowth and the invasion of other tissues. Then these cells are able to divide and grow even in the presence of signals which usually are responsible for inhibiting cell growth and division. As a result, these abnormally growing cells develop new features including changes in their cell structure, production of new biomolecules including enzymes, decreased cell adhesion thus making them gradually free to move. These heritable changes thus allow the progeny to divide and grow as their parent cell which are unaffected by the presence of normal cells that typically inhibit the growth of nearby cells. These changes allow the cancer cells to spread and invade other tissues. As mentioned, the irregularities found in cancer cells usually result from mutations in protein-encoding genes that regulate the cell division. Over a period of time more genes are self-mutated. This is so because the DNA repairing genes that encode the DNA repairing proteins are themselves not functioning normally because they are also mutated. As a result of which, mutations begin to accumulate in the cell, promoting aberrations in that cell and the progenies. Some of these mutated cells die, but other alterations may provide the abnormal cell a selective benefit that permits it to multiply much more rapidly than the normal cells. As long as these cells are restricted to their original location, they are considered as benign in nature; if they become invasive, then they are considered as malignant. Cancer cells in case of malignant tumors can often metastasize, sending cancer cells to distant sites in the body where new tumors may result.

15.3. GENETIC BASIS OF CANCER

A small fraction of the approximately 35,000 genes in the human genome have been associated with cancer. Variations in the same gene are associated with different forms of cancer thus resulting into a number of cancer types relative to the number of cancer related genes. These malfunctioning genes can be chiefly classified into three groups. The first group, called proto-oncogenes, produces protein products that usually boost cell division or inhibit normal cell death. The mutated forms of these genes are called oncogenes. The second group, called tumor suppressors, makes proteins that normally prevent cell division or cause cell death. The third group contains DNA repair genes, which help prevent mutations that lead to cancer. Proto-oncogenes and tumor suppressor genes work much like an antagonist, opposing each other. Controlled cell growth is maintained by tight regulation of proto-oncogenes, which accelerate growth, and tumor suppressor genes, which slow cell growth. Mutations that produce oncogenes accelerate growth while those that affect tumor suppressors prevent the normal inhibition of growth. Either case, leads to uncontrolled cell growth.

15.4. ONCOGENES AND SIGNAL TRANSDUCTION

In normal physiological cells, proto-oncogenes encode the proteins that send a signal to the nucleus to arouse cell division. These signaling proteins act in a series of downstream steps called as signal transduction cascade or pathway. This cascade includes a membrane receptor for the signal molecule which acts as a stimulus, intermediary proteins that carry the signal through the cytoplasm, and finally the transcription factors in the nucleus that activate the genes for cell division. In each step of the pathway, one factor or protein activates the downstream molecule ad itself gets recycled for being used in next round. Oncogenes are altered versions of the proto-oncogenes that code for these signaling molecules. The oncogenes activate the signaling cascade endlessly, ensuing an increased production of factors that stimulate growth. For example, MYC is a proto-oncogene that codes for a transcription factor. Mutations in MYC convert it into an oncogene which is roughly associated with a large variety of cancers. RAS is another oncogene that usually functions as an "on-off" switch in the signal cascade. Mutations in RAS cause the signaling pathway to remain in "on," condition thus leading to uncontrolled cell growth. About thirty percent of tumors including lung, colon, thyroid, and pancreatic carcinomas have a mutation in RAS. The renovation of a proto-oncogene to an oncogene may occur by various molecular mechanisms including mutation of the proto-oncogene, by rearrangement of genes in the chromosome that moves the proto-oncogene to a new location, or by an increase in the number of copies of the normal proto-oncogene. Occasionally a virus may insert its DNA in or near the proto-oncogene, instigating it to become an oncogene. The result of any of these events is an altered form of the gene, which contributes to cancer. Most oncogenes are dominant mutations; a single copy of this gene is enough for expression of the growth trait. This is also a "gain of function" mutation since the cells with the mutant form of the protein have gained a new function absent in cells with the normal gene. Likewise, one copy of an oncogene is sufficient to cause variations in cell growth. The occurrence of an oncogene in a germ line cell (egg or sperm) marks in an inherited predisposition for tumors in the offspring. Though, a single oncogene is not usually sufficient to cause cancer, so inheritance of an oncogene does not necessarily result in cancer.

15.5. TUMOR SUPPRESSOR GENES

The proteins produced from the tumor suppressor genes usually inhibit cell growth, preventing tumor formation. Mutations in these genes result in cells which are devoid of normal inhibition of cell growth and division. The products of tumor suppressor genes may act at the cell membrane, in the cytoplasm, or in the nucleus. Mutations in these genes result in a loss of function (that is, the ability to inhibit cell growth) thus they are usually recessive. Unlike oncogenes, in case of the tumor suppressor genes, the trait is not expressed unless both copies of the normal gene are mutated. The how is it that both genes can become mutated. In few cases, the first mutation is already present in a germ line cell (egg or sperm); thus, all the cells in the individual inherit it. Since the mutation is recessive, the trait is not expressed. In future a mutation may occur in the second copy of the gene in a somatic cell. In that cell both copies of the gene are mutated and thus the cell develops uncontrolled growth. An example of this is hereditary retinoblastoma, a serious cancer form of the retina that happens in early childhood. When one parent carries a mutation in one copy of the RB tumor suppressor gene, it is transmitted to offspring with a fifty percent probability. About ninety percent of the offspring who receive the one mutated RB gene from a parent also develop a mutation in the second copy of RB, usually very early in life. These individuals then develop retinoblastoma. Not all cases of retinoblastoma are hereditary: chances are there that they can occur by mutation of both copies of RB in the somatic cell of the individual. As retinoblasts are rapidly dividing cells and there are thousands of them, there is a high incidence of a mutation in the second copy of RB in individuals who inherited one mutated copy. This disease worries only young children because only individuals younger than about eight years old have retinoblasts. In adults, however, mutations in RB may lead to a susceptibility to several other forms of cancer. Some other cancers associated with defects in tumor suppressor genes include familial adenomatous polyposis of the colon (FPC), which results from mutations to both copies of the APC gene; hereditary breast cancer, resulting from mutations to both copies of BRCA2; and hereditary breast and ovarian cancer, resulting from mutations to both copies of BRCA1.

These cancer types advise that heredity is a significant factor in cancer but a number of cancers are irregular with no indication of a hereditary component. Cancers involving tumor suppressor genes are often hereditary because a parent may provide a germ line mutation in one copy of the gene. This may lead to a higher frequency of loss of both genes in the individual who inherits the mutated copy than in the general population. However, mutations in both copies of a tumor suppressor gene can occur in a somatic cell, so these cancers are not always hereditary. Somatic mutations that lead to loss of function of one or both copies of a tumor suppressor gene may be because of several environmental factors.

15.6. DNA REPAIR GENES

This group of genes are associated with the DNA repair processes and maintenance of chromosome structure. Abiotic environmental factors like ionizing radiation, UV light, and chemicals, can lead to the damage of DNA. Errors although a very rare event which occur during DNA replication can also lead to mutations. Gene products of DNA repair genes are involved in repairing any damage to chromosomes, thereby curtailing the frequency of mutations in the cell. But, when a DNA repair gene is mutated its product is no longer available thus, preventing DNA repair and letting further mutations to occur in the cell. These mutations can surge the frequency of a healthy cell to become cancerous. For example, a defect in a DNA repair gene called XP (Xeroderma pigmentosum) results in individuals who are very sensitive to UV light and have a high risk of suffering with a various types of skin cancer. There are seven XP genes, whose products eliminate DNA damage caused by UV light and other carcinogenic agents. Another example of a disease that is allied with loss of DNA repair is Bloom syndrome, an inherited disorder that leads to increased risk of cancer, lung disease, and diabetes. The DNA repair gene BLM, is required for maintaining the stable structure of chromosomes. Individuals with Bloom syndrome have a high frequency of chromosome breaks and interchanges, which can lead to the activation of oncogenes.

Some Genes Associated with Cancer

NAME	FUNCTION	EXAMPLES of Cancer/Diseases	TYPE of Cancer Gene
APC	regulates transcription of target genes	Familial Adenomatous Polyposis	tumor suppressor
BCL2	involved in apoptosis; stimulates angiogenesis	Leukemia; Lymphoma	oncogene
BLM	DNA repair	Bloom Syndrome	DNA repair
BRCA1	may be involved in cell cycle control	Breast, Ovarian, Prostatic, & Colonic Neoplasms	tumor suppressor
BRCA2	DNA repair	Breast & Pancreatic Neoplasms, Leukemia	tumor suppressor
HER2	tyrosine kinase; growth factor receptor	Breast, Ovarian Neoplasms	oncogene
MYC	involved in protein-protein interactions with various cellular factors	Burkitt's Lymphoma	oncogene
p16	cyclin-dependent kinase inhibitor	Leukemia; Melanoma; Multiple Myeloma;	tumor suppressor
p21	cyclin-dependent kinase inhibitor	Pancreatic Neoplasms	tumor suppressor
p53	apoptosis; transcription factor	Colorectal Neoplasms; Li-Fraumeni Syndrome	tumor suppressor
RAS	GTP-binding protein; important in signal transduction cascade	Pancreatic, Colorectal, Bladder Breast, Kidney & Lung Neoplasms; Leukemia; Melanoma	oncogene
RB	regulation of cell cycle	Retinoblastoma	tumor suppressor
SIS	growth factor	Dermatofibrosarcoma; Meningioma	oncogene
XP	DNA repair	Xeroderma pigmentosum	DNA repair

Table 15.1

15.7. CELL CYCLE

Normal healthy cells grow and divide in a tidy, well regulated fashion, in accordance with the cell cycle comprised of several check-points. Mutations in proto-oncogenes or in tumor suppressor genes allow a cell to grow and divide without the usual controls imposed by the cell cycle and thus making the healthy cell viable to cancerous features. A number of proteins are involved in controlling the timing of the events in the cell cycle, which is tightly regulated to make sure that cells divide only when needed. Any kind of damage in this regulation is the hallmark of cancer. The cyclin-dependent kinases are major control switches of the cell cycle. Each cyclin-dependent kinase forms a complex with a specific cyclin, a protein that binds and activates the cyclin-dependent kinase. The kinase portion of the complex is an enzyme that supplements a phosphate to various downstream proteins required for progression of a cell through the cycle. These added phosphates alter the structure of the protein and can activate or inactivate the protein, depending on its function. There are specific cyclin-dependent kinase/cyclin complexes at the entry points into the G1, S, and M phases of the cell cycle, as well as other factors that help prepare the cell to enter S phase and M phase. One of the important protein in the cell cycle is p53, a transcription factor that binds to the DNA, activating transcription of a protein called p21. P21 then blocks the activity of a cyclin-dependent kinase required for progression through G1. This blocking allows the cell to repair the DNA before it is replicated. If incase, the DNA damage is so extensive that it cannot be repaired, p53 triggers the cell to commit suicide. The most common mutation leading to cancer is in the gene that makes p53. Li-Fraumeni syndrome, an inherited predisposition to multiple cancers, outcomes from a germ line (egg or sperm) mutation in p53. Other protein factors that halts the cell cycle by inhibiting cyclin dependent kinases are p16 and RB. All of these proteins, including p53, are tumor suppressors. Cancer cells do not stop dividing whereas normal healthy cells are not allowed to divide continuously in an unregulated fashion. With reference to cell division, normal cells vary from cancer cells in at least four ways.

• Normal cells require external growth factors to divide. When production of these growth factors is inhibited by normal cell regulation, the cells stop dividing. But, cancer cells have lost the need for positive growth factors, so they divide irrespective of the presence of these factors. Consequently, they do not behave as part of the tissue as they have become independent cells.

• Normal cells show contact inhibition. They respond to contact with other cells by ceasing cell division. Therefore, cells can divide to fill in a gap, but they stop dividing as soon as there are enough cells to fill the gap. This characteristic is absent in cancer cells which continue to grow after they touch other cells, leading to the formation of a large mass of cells.

• Normal cells age and die, and are replaced in a controlled and orderly fashion by new cells. Apoptosis is the normal, programmed cell death. Each time the chromosome replicates, the telomeres shorten. In growing cells, the enzyme telomerase replaces these lost ends. Adult cells lack telomerase, limiting the number of times the cell can divide. Whereas, telomerase is activated in cancer cells, allowing an unlimited number of cell divisions.

• Normal cells stop to divide and die when there is DNA damage or when cell division is abnormal. Cancer cells continue to divide, even when there is a huge amount of DNA damage and when the cells are abnormal. These progeny cancer cells contain the abnormal DNA, thus as the cancer cells continue to divide they accumulate more and more damaged DNA.

15.8. CAUSE OF CANCER

Cancer can be multifactorial, meaning there is no single cause in for any one type of cancer.

- Cancer-causing substances (carcinogens): - Mutation or changes to the gene, such as damage or loss, can alter how that cell behaves. For example, a mutation may result in the production of too much proteins or that protein may not be made at all. Something that damages a cell, changing its behavior and makes it more likely to be cancerous is called a 'carcinogen'.
- Age: - Many types of cancer become more prevalent with age. The longer the people live, the more exposure there is to carcinogens and the more time there is for genetic changes or mutations to occur within their cells thus increasing the risk of cancer.
- Genetics: - Prevailing model for cancer development reveals that mutations in genes for tumor suppressors and oncogenes lead to cancer. However, it may not be that too simple, as the prevailing model fails to explain the genetic diversity among cells within a single tumor and does not adequately explain many chromosomal aberrations typical of cancer cells. An alternate model proposes that there are "master genes" controlling cell division. A mutation in a master gene leads to abnormal replication of chromosomes, triggering whole sections of chromosomes to be missing or duplicated. This leads to a change in gene dosage, so cells produce too little or too much of a specific protein. If the chromosomal aberrations affect the amount of one or more proteins controlling the cell cycle, such as growth factors or tumor suppressors, the result may be cancer. There is also strong evidence that the excessive addition of methyl groups to genes involved in the cell cycle, DNA repair, and apoptosis is characteristic of some cancers.
- The immune system: - People who have weakened immune systems are more disposed to developing some types of cancer. This includes people who have had organ transplants and take drugs to suppress their immune systems to stop organ rejection, people who have HIV or AIDS, or other medical conditions which reduce their immunity. Certain lifestyles and environmental factors also can cause mutations that can lead to cancer. Lifestyle and environmental causes are to a large extent controllable or avoidable. Examples include:
- Body weight, diet and physical activity - Experts estimate that maintaining a healthy bodyweight, making changes to our diet and doing regular physical activity could prevent cancer. Many people consume too much red and processed meat and not enough fresh fruit and vegetables. This type of diet is known to increase the risk of cancer

- Overweight or obesity - Overweight or obese people have a higher risk of bowel and pancreatic cancer, probably due to a tendency towards higher insulin levels. Obesity can also increase the risk of cancer of the oesophagus (oesophageal cancer), kidney and gallbladder cancer, as well as breast or womb (uterine) cancer in women.
- Alcohol - The evidence that all types of alcoholic drinks are a cause of a number of cancers is now stronger. Alcohol can increase the risk of a number of cancers, including mouth, throat (like pharyngeal cancer), laryngeal and cancer of the food pipe, liver cancer.
- Tobacco – Tobacco smoke contains at least 80 different carcinogenic agents. When smoke is inhaled the chemicals enter the lungs, pass into the blood stream and are transported throughout the body. Thus smoking or chewing tobacco not only causes lung cancer and mouth cancers, but is also related to many other cancers.
- Ionising radiation – Manmade sources of radiation can cause cancer and are a risk for people associated with it. The main risk is however, extended and unprotected exposure to ultraviolet radiations from the sun which can lead to melanoma and skin malignancies. Fair skinned people, those with lot of moles or who have a relative who has had melanoma or non-melanoma skin cancer, are at utmost risk. Curie, who discovered radium, paving the way for radiation therapy for cancer, died of cancer herself as a result of radiation exposure in her research.
- Work place hazards - Some people risk being exposed to a cancer causing substance because of the work they are associated with. Workers in the chemical dye industry have been found to have a higher incidence than normal for bladder cancer. Asbestos is a well-known work place cause of cancer, particularly a cancer called mesothelioma, which most commonly affects the covering of the lungs (pleura).
- Infection – Some cancers can also be caused by infection with a virus. The virus can cause changes in cells that make them more likely to become cancerous. Examples include cervical cancer, linked to the Human *Papilloma Virus*, primary liver cancer which can be caused by the Hepatitis B and C virus and lymphomas linked to the *Epstein-Barr virus*. Previously, bacterial infections have not been thought of as cancer causing agents. But studies have shown that people who have helicobacter pylori infection of their stomach, develop inflammation of the stomach lining, which increases the risk of stomach cancer.
- There may be multiple mechanisms leading to the development of cancer. This, further complicates the difficult task of determining what causes cancer.

15.9. TUMOR BIOLOGY:

Cancer cells act as independent cells, growing in an uncontrollable fashion to form tumors. Tumors grow in a series of steps. The first step is hyperplasia, meaning that there are too many cells resulting from uncontrolled cell division. These cells look normal, but changes have occurred that result in some loss of control of growth. The second step is dysplasia, resulting from further growth, accompanied by abnormal changes to the cells. The third step requires additional changes, which result in cells that are even more abnormal and can now spread over a wider area of tissue. These cells begin to lose their original function; such cells are called anaplastic. At this stage, since the tumor is still restricted within its original location (called in situ) and is not invasive, it is not considered malignant but it is potentially malignant. The last step occurs when the cells in the tumor metastasize, which means that they can invade surrounding tissue, including the bloodstream, and spread to other locations. This is the most serious type of tumor, but not all tumors progress to this point. Non-invasive tumors are said to be benign. The type of tumor that forms depends on the type of cell that was initial site of infection. There are five types of tumors.

- Carcinomas result from altered epithelial cells. This arises from the epithelial cells (the lining of cells that helps protect or enclose organs). Carcinomas may invade the surrounding tissues and organs and metastasize to the lymph nodes and other areas of the body. The most common forms of cancer in this group are breast, prostate, lung and colon cancer.
- Sarcomas result from changes in muscle, bone, fat, or connective tissue. A type of malignant tumor of the bone or soft tissue (fat, muscle, blood vessels, nerves and other connective tissues that support and surround organs). The most common forms of sarcoma are leiomyosarcoma, liposarcoma and osteosarcoma.
- Leukemia results from malignant white blood cells. Leukaemia is a cancer of the white blood cells and bone marrow, the tissue that forms blood cells. There are several subtypes; common are lymphocytic leukaemia and chronic lymphocytic leukaemia.
- Lymphoma is a cancer of the lymphatic system cells that derive from bone marrow. Lymphoma is a cancer of the lymphatic system, which runs all through the body, and can therefore occur anywhere. The two main forms are non-Hodgkin's which begins with uncontrolled growth of the - white blood cells -lymphocytes - of the immune system) and Hodgkin's lymphoma in which cells of the lymph nodes become cancerous.
- Myelomas are cancers of specialized white blood cells that make antibodies.

Angiogenesis:

Any cell whether healthy or cancerous requires nutrients and oxygen in order to grow. All living tissues are adequately supplied with capillary vessels which circulate nutrients and oxygen to every cell. As tumors expand, the cells in the center no longer receive nutrients from the normal blood vessels. To provide a blood supply for all the cells in the tumor, it must form new blood vessels to supply the cells with nutrients and oxygen. Angiogenesis, is a process where tumor cells make growth factors which induce formation of new capillary blood vessels. The cells of the blood vessels that divide to make new capillary vessels are inactive in normal tissue; however, tumors can make angiogenic factors, which activate these blood vessel cells to divide. Without the additional blood supplied by angiogenesis, tumors cannot grow larger and also cannot spread, or metastasize to new tissues. Tumor cells can cross through the walls of the capillary blood vessel at a rate of about one million cells per day. Still not all cells in a tumor are angiogenic. Both angiogenic and non-angiogenic cells in a tumor cross into blood vessels and spread. However, non-angiogenic cells give rise to dormant tumors when they grow in other locations. In contrast, the angiogenic cells quickly establish themselves in new locations by growing and producing new blood vessels, resulting in rapid growth of the tumor. Tumors produce angiogenic factors. An oncogene called BCL2 has been shown to greatly surge the production of a potent stimulator of angiogenesis. There are several angiogenic factors and production of many of these is increased by a variety of oncogenes. Thus, oncogenes in some tumor cells allow those cells to produce angiogenic factors. The progeny of these tumor cells will also produce angiogenic factors, so the population of angiogenic cells will increase as the size of the tumor increases. Angiogenesis is critical for the progression of dormant tumors into cancer.

15.10. SIGNS AND SYMPTOMS

As there are so many different types of cancer, the symptoms are varied and depend on where the disease is located. However, there are some key signs and symptoms, including:

• Lumps: - some cancers can be felt through the skin. Cancerous lumps are often painless and may increase in size as the cancer progresses.

• Coughing, breathlessness: - persistent coughing episodes and breathlessness can be associated with lung cancer.

• Changes in bowel habit: - symptoms of bowel cancer may include blood in the stools and a change in bowel habits such as constipation and diarrhea.

• Bleeding: - any unexpected bleeding can be a sign of cancer:

o Bleeding from the anal passage may be a sign of bowel cancer.

o Bleeding from the cervix may be a sign of cervical cancer.

o Blood present in the urine may be a sign of kidney or bladder cancer.

• Unexplained weight loss: - a large amount of unexplained weight loss over a short period of time (a couple of months) can be a sign of cancer.

• Fatigue: - fatigue is extreme tiredness and a severe lack of energy. If fatigue is due to cancer, sufferers normally also have other symptoms.

15.11. DETECTING AND DIAGNOSING CANCER

Imaging techniques such as MRI, X-rays, CT, and ultrasound are the most common techniques which can provide an image of a tumor. Endoscopy allows to look for tumors in organs such as the stomach, colon, and lungs. Most of these techniques are used to detect visible tumors, which must then be removed by biopsy and examined microscopically by a pathologist. Then the aberrations in the cells in terms of their shape, size, and structure, especially the nucleus need to be sorted out. Based on investigation of the tumor cells, the tumor can be classified as benign or malignant, and also the stage of development of tumor can be defined whether in early or late stage. Tumor markers proteins are found more frequently in the blood of individuals with the tumor than in normal individuals. These are not ideal compounds for diagnosing of cancer for two reasons. First, individuals without cancer may have elevated levels of the marker, leading to false positives. Second, tumor markers are not sufficiently elevated in all individuals with cancer to allow their detection. This leads to false negatives. One of the most commonly used tumor markers is prostate-specific antigen (PSA). It is present in all adult males, but its level is amplified after both benign and malignant changes in the prostate. Thus, high levels of PSA indicate only that further tests are prerequisite to determine whether the condition is cancer. If prostate cancer is diagnosed, the levels of PSA can help to determine the effectiveness of treatment and detect recurrence. CA125 is another tumor marker, which is produced by a number of different cells, particularly ovarian cancer cells. It is used primarily to monitor the treatment efficacy of ovarian cancer. When the cancer is responding to treatment, CA125 levels drop. It is not used as a routine test for ovarian cancer as many common conditions that cause inflammation also increase the level of CA125, leading to a high frequency of false positives. The earlier a cancer is found the more effectively it can be treated; however, early stage cancers typically produce no symptoms. Scientists are evolving molecular techniques to detect very early cancer. Using techniques such as mass spectrometry, they are also developing specific blood tests to identify a pattern of new proteins in the blood of individuals with a particular type of cancer. Scientists are also developing DNA microarrays to identify genes expressed in particular types of cancer cells. With the sequencing of the human genome and the mapping of single nucleotide polymorphisms (SNPs), it may be possible to diagnose particular cancers by identifying cells with known gene alterations. In 2002 scientists detected ovarian cancer by testing blood for the presence of DNA released by tumor cells. They looked for changes in certain alleles at eight SNPs that are characteristic of cancer. By using this technique, they could successfully identified eighty-seven percent of patients known to have early-stage of ovarian cancer and ninety-five percent of those with late-stage ovarian cancer. The capability to determine which genetic alterations are related with various cancers unlocks the possibility of recognizing cancerous cells while the cancer is in an early, treatable stage.

15.12. TREATMENT

Treatment of cancer depends on the type of cancer, its location, and its state of development. Surgery is used to remove solid tumors. This treatment may be necessary for early stage cancers and benign tumors. Radiation treatment destroys cancer cells with high-energy rays targeted directly to the tumor site. It acts mainly by damaging DNA and preventing its replication. Therefore, it preferentially kills cancer cells. It also kills some normal cells, predominantly those that are dividing. Surgery and radiation treatment are often used together. Chemotherapy drugs are toxic compounds that target quickly growing cells. Many of these drugs are specially designed to restrict the synthesis of precursor molecules needed for DNA replication, they interfere with the ability of the cell to complete the S phase of the cell cycle. Other drugs cause extensive DNA damage, which stops replication. A class of drugs known as spindle inhibitors stops cell replication early in mitosis. During the mitosis, chromosome separation requires spindle fibers composed of microtubules. Spindle inhibitors stop the synthesis of microtubules. Since, most adult cells don't divide often, they are less sensitive to these drugs than are cancer cells. Chemotherapy drugs also destroy certain adult cells that divide more rapidly, like those that line the gastrointestinal tract, bone marrow cells, and hair follicles. This results in some of the side effects of chemotherapy, including gastrointestinal distress, low white blood cell count, and hair loss. Although cancer cells have lost some of the normal responses to growth factors, some cancer cells still require hormones for growth. Hormone therapy for cancer tries to starve the cancer cells of these hormones. This is typically done by drugs that block the activity of the hormone, although some drugs can block synthesis of the hormone. For example, some breast cancer cells require estrogen for growth. Drugs that block the binding site for estrogen can slow the growth of these cancers. These drugs are called selective estrogen receptor modulators (SERMs) or anti-estrogens. Tamoxifen and Raloxifene are examples of this type of drug. Likewise, testosterone (an androgen hormone) stimulates some prostate cancer cells. Selective androgen receptor modulators (SARMs) are drugs that block the binding of testosterone to these cancer cells, inhibiting their growth and possibly preventing prostate cancer. Newer chemotherapeutic drugs target specific, active proteins or processes in cancer cell signal transduction pathways, such as receptors, growth factors, or kinases. As the targets are cancer-specific proteins, the faith is that these drugs will be much less toxic to normal cells than conventional cancer drugs. Chemotherapy may fail because the cancer cells become resistant to the therapeutic drugs. One of the characteristics of cancer cells is a high frequency of mutation. In the presence of toxic drugs, cancer cells that mutate and become resistant to the drug will survive and multiply in the presence of the drug, producing a tumor that is also resistant to the drug. So many a times, combinations of chemotherapy drugs are given at the same time. This decreases the probability that a cell will develop resistance to several drugs at once. However, such multiple resistances do occur. Some drug-resistant cancer cells express a gene called MDR1 (multiple drug resistance). This gene encodes a membrane protein that can not only prevent some drugs from entering the cell, but can also expel drugs already in the cell. Another hopeful target for cancer therapy is angiogenesis. Numerous drugs, including some naturally occurring compounds, have the ability to inhibit angiogenesis. Two compounds in this class are angiostatin and endostatin, both are derived from naturally occurring proteins. These drugs prevent angiogenesis by tumor cells, restricting tumor growth and preventing metastasis. Advantage of angiogenesis inhibitors is that, because they do not target the cancer cells directly, there is less chance that the cancer cells will develop resistance to the drug. Immunotherapy includes several techniques that use the immune system to attack cancer cells or treat the side effects of some types of cancer treatment. A technique called chemoimmunotherapy attaches chemotherapy drugs to antibodies that are specific for cancer cells. The antibody then carries the drug directly to cancer cells without harming normal cells, reducing the toxic side effects of chemotherapy. A similar strategy, radioimmunotherapy, couples specific antibodies to radioactive atoms, thereby targeting the deadly radiation specifically to cancer cells.

Some Drugs Used in the Treatment of Cancer

CLASS	MECHANISM
selective estrogen receptor modulators (SERM); (Tamoxifen and Raloxifene)	blocks the binding site for estrogen; can slow the growth of estrogen-stimulated cancers
selective androgen receptor modulators (SARM)	blocks the binding site for testosterone; can slow the growth of testosterone-stimulated cancers
spindle inhibitors	stops cell replication early in mitosis
farnesyl transferase inhibitors	blocks the addition of a farnesyl group to RAS, preventing its activation
Gleevec®	binds to abnormal proteins in cancer cells, blocking their action
angiogenesis inhibitors (endostatin, angiostatin)	prevent angiogenesis by tumor cells
immunostimulants (interleukin 2, alpha interferon)	enhance the normal immune response
Herceptin®	antibody that binds to HER2 receptor on tumor cells, preventing the binding of growth factors

Table 15.2

Cancer seems to result from a mixture of genetic changes and environmental factors. Lifestyle that minimizes exposure to environmental carcinogens is one effective means of preventing cancer. Individuals who restrict their exposure to tobacco products, sunlight, and pollution can greatly decrease their risk of developing cancer. Many foods contain antioxidants and other nutrients that may help to prevent cancer. The National Cancer Institute recommends a diet with large amounts of colorful fruits and vegetables. These foods supply sufficient amounts of vitamin A, C, and E, as well as phytochemicals and other antioxidants that help to prevent cancer. Vaccines are also potential candidates for prevention of cancer. It appears that vaccines such as hepatitis B vaccine, papillomavirus vaccine may play a crucial role in minimizing the risk of liver cancer and cervical cancer (precancerous lesions) respectively.

By here and now, we discern that cancer is not simply the localized lumps and bumps that we have been programmed to accept through the years. Cancer in the adult can often be seen as a deteriorating process with symptoms representative of underlying systemic dysfunction. There are countless factors, including emotional, diet, drugs and chemicals, infections, genetic mutation and environmental pollutants. Conventional treatments look at cancer as a disease state. The natural-oriented doctor views cancer as a set of symptoms reflecting underlying disease. The conventional treatment of surgery, radiation and chemotherapy has been the cornerstone of cancer treatment over the past several decades. Less toxic and target-specific chemotherapeutic agents are being developed. Further research and clinical studies are also being conducted on natural therapies. The success of a treatment often depends on the stage of cancer, the age, the immunity status, and the tumor response rate of the patient. As more research is carried out new therapies will be found. Regrettably, most cancer patients do not have time to wait. Sometimes, they are only given a few more months to live and left behind with an unsolved question Why did this happen to me?

REFERENCES:

- Schneider, K. 2001. Counseling about cancer: Strategies for genetic counseling. 2d ed. New York: John Wiley & Sons.
- Bahls, C., and M. Fogarty. 2002. Reining in a killer disease. The Scientist 16[11]:16. An outline of several approaches to controlling cancer.
- Gibbs, W. Wayt. 2003. Untangling the roots of cancer. Scientific American, July, 57–65. New evidence challenges old theories of how cancer develops.
- Kling, Jim. 2003. Put the blame on methylation. The Scientist
- 16[12]:27–28. Gene silencing by methylation rather than by gene mutation may create some cancer cells.
- McCook, A. 2002. Lifting the screen. Scientific American, June, 16–17. Developing protein "fingerprints" to screen for cancer.

- Rayl, A. J. S., and Lewis, R. 2001. In cancer research, diet and exercise roles strengthen. The Scientist 15[20]:17. Evidence for the importance of lifestyle in preventing cancer.
- Veggeberg, S. 2002. Fighting cancer with angiogenesis inhibitors. The Scientist 16[11]:41. Discussion of a class of drugs that helps to prevent angiogenesis.
- Wilson, J. F. 2001. A dual role for CDK inhibitors. The Scientist 16[6]:20. Discusses approaches to cancer treatment using cells' cycle inhibitors.
- Wilson, J. F. 2002. Elucidating the DNA damage pathway. The Scientist 16[2]:30. How researchers have deciphered the role of DNA damage repair in cancer.

16. JAUNDICE

G.K. Panigrahi

16.1. INTRODUCTION

Jaundice, also known as icterus, is a yellowish or greenish pigmentation of the skin and whites of the eyes due to high bilirubin levels. It is commonly associated with itchiness. The faeces may be pale and the urine dark. Jaundice in babies occurs in over half in the first week following birth and in most is not a problem. If bilirubin levels in babies are very high for too long a type of brain damage, known as kernicterus, may occur.

Causes of jaundice vary from non-serious to potentially fatal. Levels of bilirubin in blood are normally below 1.0 mg/dL (17 µmol/L) and levels over 2–3 mg/dL (34-51 µmol/L) typically results in jaundice. High bilirubin is divided into two types: unconjugated (indirect) and conjugated (direct). Conjugated bilirubin can be confirmed by finding bilirubin in the urine. Other conditions that can cause yellowish skin but are not jaundice include carotenemia from eating large amounts of certain foods and medications like rifampin.

High unconjugated bilirubin may be due to excess red blood cell breakdown, large bruises, genetic conditions such as Gilbert's syndrome, no eating for a prolonged period of time, newborn jaundice, or thyroid problems. High conjugated bilirubin may be due to liver diseases such as cirrhosis or hepatitis, infections, medications, or blockage of the bile duct. In the developed world the cause is more often blockage of the bile duct or medications while in the developing world it is more often infections such as viral hepatitis, leptospirosis, schistosomiasis, or malaria. Blockage of the bile duct may occur due to gallstones, cancer, or pancreatitis. Medical imaging such as ultrasound is useful for detecting bile duct blockage.

Treatment of jaundice is typically determined by the underlying cause. If a bile duct blockage is present surgery is typically required, otherwise management is medical. Medical management may involve treating infectious causes and stopping medication that could be contributing. Among newborns, depending on age and prematurity, a bilirubin greater than 4-21 mg/dL (68-360 µmol/L) may be treated with phototherapy or exchanged transfusion. The itchiness may be helped by draining the gallbladder or ursodeoxycholic acid. The word jaundice is from the French jaunisse, meaning "yellow disease".

16.2. SIGNS AND SYMPTOMS

The main symptom of jaundice is a yellowish discoloration of the white area of the eye and the skin. Urine is dark in colour. Slight increases in serum bilirubin are best detected by examining the sclerae, which have a particular affinity for bilirubin due to their high elastin content. The presence of scleral icterus indicates a serum bilirubin of at least 3 mg/dL. The conjunctiva of the eye are one of the first tissues to change color as bilirubin levels rise in jaundice. This is sometimes referred to as scleral icterus. However, the sclera themselves are not "icteric" (stained with bile pigment) but rather the conjunctival membranes that overlie them. The yellowing of the "white of the eye" is thus more properly termed conjunctival icterus. The term "icterus" itself is sometimes incorrectly used to refer to jaundice that is noted in the sclera of the eyes, however its more common and more correct meaning is entirely synonymous with jaundice.

16.3. TYPES OF JAUNDICE

When a pathological process interferes with the normal functioning of the metabolism and excretion of bilirubin just described, jaundice may be the result. Jaundice is classified into three categories, depending on which part of the physiological mechanism the pathology affects. The three categories are:

Category	Definition
Pre-hepatic/ hemolytic	The pathology is occurring prior to the liver due to either: A. Intrinsic defects in RB cells B. Extrinsic causes external to RB cells
Hepatic/ hepatocellular	The pathology is located within the liver caused due to disease of parenchymal cells of liver.
Post-Hepatic/ cholestatic	The pathology is located after the conjugation of bilirubin in the liver caused due to obstruction of biliary passage.

Table 16.1

16.3.1. Pre-hepatic

Pre-hepaticular jaundice is caused by anything which causes an increased rate of haemolysis (breakdown of red blood cells). Unconjugated bilirubin comes from the breakdown of the haeme pigment found in red blood cells' haemoglobin. The increased breakdown of red blood cells leads to an increase in the amount of unconjugated bilirubin present in the blood and deposition of this unconjugated bilirubin into various tissues can lead to a jaundiced appearance. In tropical countries, severe malaria can cause jaundice in this manner. Certain genetic diseases, such as sickle cell anaemia, spherocytosis, thalassemia, pyruvate kinase deficiency, and glucose 6-phosphate dehydrogenase deficiency can lead to increased red cell lysis and therefore haemolytic jaundice. Commonly, diseases of the kidney, such as haemolytic uremic syndrome, can also lead to coloration. Defects in bilirubin metabolism also leads to jaundice, as in Gilbert's syndrome (a genetic disorder of bilirubin metabolism which can result in mild jaundice, which is found in about 5% of the population) and Crigler-Najjar syndrome, Type I and II.

In jaundice secondary to haemolysis, the increased production of bilirubin leads to the increased production of urine-urobilinogen. Bilirubin is not usually found in the urine because unconjugated bilirubin is not water-soluble, so, the combination of increased urine-urobilinogen with no bilirubin (since, unconjugated) in urine is suggestive of hemolytic jaundice. Laboratory findings include:

Urine: no bilirubin present, urobilinogen > 2 units (i.e., haemolytic anemia causes increased haeme metabolism; exception: infants where gut flora has not developed).

Serum: increased unconjugated bilirubin.

Kernicterus is associated with increased unconjugated bilirubin; neonates are especially vulnerable to this due to increased permeability of the blood brain barrier.

16.3.2. Hepatocellular

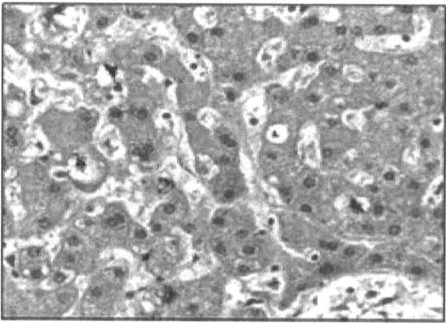

Figure 16.1

Microscopy of cholestatic liver showing bilirubin pigment
(Source-Wikipedia)

Hepatocellular (hepatic) jaundice can be caused by acute or chronic hepatitis, hepatotoxicity, cirrhosis, drug-induced hepatitis and alcoholic liver disease. Cell necrosis reduces the liver's ability to metabolize and excrete bilirubin leading to a buildup of unconjugated bilirubin in the blood. Other causes include primary biliary cirrhosis leading to an increase in plasma conjugated bilirubin because there is impairment of excretion of conjugated bilirubin into the bile. The blood contains an abnormally raised amount of conjugated bilirubin and bile salts which are excreted in the urine. Jaundice seen in the newborn, known as neonatal jaundice, is common in newborns as hepatic machinery for the conjugation and excretion of bilirubin does not fully mature until approximately two weeks of age. Rat fever (leptospirosis) can also cause hepatic jaundice. In hepatic jaundice, there is invariably cholestasis.

16.4. SYMPTOMS

Urine: Conjugated bilirubin present, urobilirubin > 2 units but variable (except in children). Kernicterus is a condition not associated with increased conjugated bilirubin.

Plasma protein show characteristic changes.

Plasma albumin level is low but plasma globulins are raised due to an increased formation of antibodies.

Bilirubin transport across the hepatocyte may be impaired at any point between the uptake of unconjugated bilirubin into the cell and transport of conjugated bilirubin into biliary canaliculi. In addition, swelling of cells and oedema due to inflammation cause mechanical obstruction of intrahepatic biliary tree. Hence in hepatocellular jaundice, concentration of both unconjugated and conjugated bilirubin rises in the blood. In hepatocellular disease, there is usually interference in all major steps of bilirubin metabolism—uptake, conjugation and excretion. However, excretion is the rate-limiting step, and usually impaired to the greatest extent. As a result, conjugated hyperbilirubinaemia predominates.

The unconjugated bilirubin still enters the liver cells and becomes conjugated in the usual way. This conjugated bilirubin is then returned to the blood, probably by rupture of the congested bile canaliculi and direct emptying of the bile into the lymph leaving the liver. Thus, most of the bilirubin in the plasma becomes the conjugated type rather than the unconjugated type, and this conjugated bilirubin which did not go to intestine to become urobilinogen gives the urine the dark color.

16.4.1. Post-hepatic

Post-hepatic jaundice, also called obstructive jaundice, is caused by an interruption to the drainage of bile containing conjugated bilirubin in the biliary system. The most common causes are gallstones in the common bile duct, and pancreatic cancer in the head of the pancreas. Also, a group of parasites known as "liver flukes" can live in the common bile duct, causing obstructive jaundice. Other causes include strictures of the common bile duct, biliary atresia, cholangiocarcinoma, pancreatitis, cholestasis of pregnancy, and pancreatic pseudocysts. A rare cause of obstructive jaundice is Mirizzi's syndrome. In complete obstruction of the bile duct, no urobilinogen is found in the urine, since bilirubin has no access to the intestine and it is in the intestine that bilirubin gets converted to urobilinogen to be later released into the general circulation. In this case, presence of bilirubin (conjugated) in the urine without urine-urobilinogen suggests obstructive jaundice, either intra-hepatic or post-hepatic. The presence of pale stools and dark urine suggests an obstructive or post-hepatic cause as normal feces get their color from bile pigments. However, although pale stools and dark urine are a feature of biliary obstruction, they can occur in many intra-hepatic illnesses and are therefore not a reliable clinical feature to distinguish obstruction from hepatic causes of jaundice.

Patients also can present with elevated serum cholesterol, and often complain of severe itching or "pruritus" because of the deposition of bile salts. No single test can differentiate between various classifications of jaundice. A combination of liver function tests is essential to arrive at a diagnosis.

16.4.2. Neonatal jaundice

Neonatal jaundice is usually harmless: this condition is often seen in infants around the second day after birth, lasting until day 8 in normal births, or to around day 14 in premature births. Typical causes for neonatal jaundice include normal physiologic jaundice, jaundice due to formula supplementation, and hemolytic disorders that include hereditary spherocytosis, glucose-6-phosphate dehydrogenase deficiency, pyruvate kinase deficiency, ABO/Rh blood type autoantibodies, or infantile pyknocytosis. Serum bilirubin normally drops to a low level without any intervention required. In cases where bilirubin rises higher, a brain-damaging condition known as kernicterus can occur, leading to significant disability. This condition has been rising in recent years due to less time spent outdoors. A Bili light is often the tool used for early treatment, which often consists of exposing the baby to intensive phototherapy. Sunbathing is effective treatment, and has the advantage of ultra-violet-B, which promotes Vitamin D production. Bilirubin count is lowered through bowel movements and urination, so frequent and effective feedings are especially important.

16.5. DIFFERENTIAL DIAGNOSIS

Yellow discoloration of the skin, especially on the palms and the soles, but not of the sclera or inside the mouth is due to carotenemia—a harmless condition.

16.6. PATHOPHYSIOLOGY

Jaundice itself is not a disease, but rather a sign of one of many possible underlying pathological processes that occur at some point along the normal physiological pathway of the metabolism of bilirubin in blood.

When red blood cells have completed their life span of approximately 120 days, or when they are damaged, their membranes become fragile and prone to rupture. As each red blood cell traverses through the reticulo-endothelial system, its cell membrane ruptures when its membrane is fragile enough to allow this. Cellular contents, including hemoglobin, are subsequently released into the blood. The hemoglobin is phagocytosed by macrophages, and split into its heme and globin portions. The globin portion, a protein, is degraded into amino acids and plays no role in jaundice. Two reactions then take place with the haeme molecule. The first oxidation reaction is catalyzed by the microsomal enzyme haeme oxygenase and results in biliverdin (green color pigment), iron and carbon monoxide. The next step is the reduction of biliverdin to a yellow color tetrapyrol pigment called bilirubin by cytosolic enzyme biliverdin reductase. This bilirubin is "unconjugated," "free" or "indirect" bilirubin. Approximately 4 mg of bilirubin per kg of blood is produced each day. The majority of this bilirubin comes from the breakdown of haeme from expired red blood cells in the process just described. However approximately 20 percent comes from other haeme sources, including ineffective erythropoiesis, and the breakdown of other haeme-containing proteins, such as muscle myoglobin and cytochromes.

16.7. HEPATIC EVENTS

The unconjugated bilirubin then travels to the liver through the bloodstream. Because this bilirubin is not soluble, however, it is transported through the blood bound to serum albumin. Once it arrives at the liver, it is conjugated with glucuronic acid (to form bilirubin diglucuronide, or just "conjugated bilirubin") to become more water-soluble. The reaction is catalyzed by the enzyme UDP-glucuronyl transferase.

This conjugated bilirubin is excreted from the liver into the biliary and cystic ducts as part of bile. Intestinal bacteria convert the bilirubin into urobilinogen. From here urobilinogen can take two pathways. It can either be further converted into stercobilinogen, which is then oxidized to stercobilin and passed out in the faeces, or it can be reabsorbed by the intestinal cells, transported in the blood to the kidneys, and passed out in the urine as the oxidised product urobilin. Stercobilin and urobilin are the products responsible for the coloration of feces and urine, respectively.

16.8. EPIDEMIOLOGY

It is unclear how common it is among adults.

16.8.1. DIAGNOSTIC APPROACH

Most patients presenting with jaundice will have various predictable patterns of liver panel abnormalities, though significant variation does exist. The typical liver panel will include blood levels of enzymes found primarily from the liver, such as the aminotransferases (ALT, AST), and alkaline phosphatase (ALP); bilirubin (which causes the jaundice); and protein levels, specifically, total protein and albumin. Other primary lab tests for liver function include gamma glutamyl transpeptidase (GGT) and prothrombin time (PT).

REFERENCES

- Bassari, R; Koea, JB (7 February 2015). "Jaundice associated pruritis: a review of pathophysiology and treatment.". World journal of gastroenterology. 21 (5): 1404–13.
- Bertini, G.; Dani, C.; Tronchin, M.; Rubaltelli, F. F. (2001). "Is Breastfeeding Really Favoring Early Neonatal Jaundice?". Pediatrics. 107 (3): E41.
- Click, Rachel; Dahl-Smith, Julie; Fowler, Lindsay; DuBose, Jacqueline; Deneau-Saxton, Margi; Herbert, Jennifer (2013). "An osteopathic approach to reduction of readmissions for neonatal jaundice". Osteopathic Family Physician. 5 (1): 17–23.

- Nakayama, Juichiro; Imafuku, Shinichi; Mori, Tatsuki; Sato, Chiemi (2013). "Narrowband ultraviolet B irradiation increases the serum level of vitamin D3 in patients with neurofibromatosis 1". The Journal of Dermatology. 40 (10): 829–31.
- Roche, SP; Kobos, R (15 January 2004). "Jaundice in the adult patient.". American family physician. 69 (2): 299–304.
- Salih, Fadhil M. (2001). "Can sunlight replace phototherapy units in the treatment of neonatal jaundice? An in vitro study". Photodermatology, Photoimmunology and Photomedicine. 17 (6): 272–7.
- Shinde, MN Chatterjea, Rana (2012). Textbook of medical biochemistry (8th ed.). New Delhi: Jaypee Brothers Medical Publications (P) Ltd. p. 672. ISBN 978-93-5025-484-4.
- 8.Winger, J; Michelfelder, A (September 2011). "Diagnostic approach to the patient with jaundice.". Primary care. 38 (3): 469–82; viii.

SOURCE OF FIGURES AND TABLES

Source of figures and tables have been mentioned below each figure in the text and mentioned under "references" at the end of each chapter.

DISEASES AT A GLANCE

Sl.	Diseases	Cause	Symptoms	Risk factor	Prevention	Treatment
1	Measles	Measles Virus/ *Rubeola virus*	Fever, cold, cough, sneezing, Conjuctivities, rashes	Contact with mucus and saliva of infected person	Measles Vaccine	Supportive care, rest and intake of fluids to prevent dehydration
2	Smallpox	*Variola major, Variola minor*	Influenza, Common cold, Fever, Muscle pain, Headache, nausea, vomiting, backache	Coughing, sneezing or direct contact with body fluids of patients	Smallpox vaccine	Supportive care, antiviral drugs
3	Chickenpox	*Varicella zoster virus*	Small, itchy blisters, headache	Contact with blisters, coughs and sneezes	Varicella vaccine	Calamine lotion, paracetamol, Aciclovir
4	AIDS	*HIV*	Flu like illness, large lymph nodes, fever, weight loss	Exposure to blood, breast milk, Sex	Safe sex, needle exchange, male circumcision	Antiretroviral thepay
5	Ebola	*Ebola virus*	Sudden fever, headache, weakness, muscle pain, vomiting	Avoid contact with blood and body fluids	Careful hygiene, Avoid contact with patients	Antiviral drug
6	Dengue	*Dengue virus*	Dangue fever, Dengue haemorrhagic fever, muscle and joint pain	Aedes aegypti mosquito, Environment	Control of mosquito and environment management	Paracetamol preparation drugs
7	Cholera	*Vibrio cholerae*	Diarrhoea, vomiting, muscle cramps	Poor sanitation, Not enough clean drinking water, poverty	Clean water, Cholera vaccine	Oral rehydration therapy, Zinc supplementation, Intravenous fluids, Antibiotics
8	Leprosy	*Mycobacterium leprae*	Decrease ability to feel pain	Close contact with a case of leprosy, poverty	Avoid Cough or contact with fluid from the nose of an infected person	Multidrug therapy
9	TB	*Mycobacterium tuberculosis, M. bovis*	Cough with blood or sputum, pain in the chest, weakness, weight loss, no appetite, fever	Through air from person to person	Less exposure to patient	Antibiotics

10	Typhoid	*Salmonella enterica serotype typhi*	Fever,dry cough,abdominal discomfort	Contaminated water,close contact with patients,poor housing and inadequate hygiene	Avoid contact with patients,contaminated water	Oral or intravenous hydration,Use of antipyretics, appropriate nutrition and blood transfusion
11	Malaria	*Plasmodium parasites*	Fever,headache,vomiting,fatigue, anaemia	Climatic condition,mosquitoes	Avoid being bitten by mosquitoes	Antimalarial drugs
12	Amoebiasis	*Entamoeba histolytica*	Diarrhoea with blood and mucus	Contaminated water	Hygiene,Avoid eating street foods,raw vegetables	Antiamoebic drugs
13	Diarrhoea	Viral, bacterial, parasitic	Loose bowel movement,dehydration	Contaminated food ,water	Hand washing, rota virus vaccination,breast feeding	ORS,Zinc supplementation
14	Filariasis	*Wuchereria bancrofti*				
15	Cancer	Multifactors and oncogenes	Lumps,coughing,breathlessness,change in bowel habit	Chemicals,virus ,ionising radiation,tobacco,alcohol,over weight	Control of weight, Avoid carcinogens	Surgery, radiation, chemotherapy
16	Jaundice	High bilirubin in blood	Yellowish skin,eye ball,stool & urine,itching	Contaminated water	Good sanitary habit	Surgery and medication

Table 16.2

ABOUT THE EDITORS

Sasmita Panda born in 1982 and graduated in the year 2002 from Godavaris Mahavidyalaya, Banpur under Utkal University, Odisha. She completed P.G. in Zoology in the year 2004 and M. Phil. In the year 2013 from Khallikote college under Berhampur University .She has been teaching Zoology in different U.G. Colleges since 2006. Now serving as lecturer in zoology at Jatni college, Jatni. She has published many articles in research journals of National and International repute. She has also co-authored a Fishery book entitled "Employment Through Aquaculture", "Biology of Wild Animals", "Wild Animals of India" and "Earning Animals" meant for students of undergraduate and post-graduate level. She has also edited a book entitled "Water for Survival".

Dr. S.N. Padhi born in 1953 obtained B.Sc. degree in 1973, P.G. in 1975 and Ph.D. in the year 1981 under Berhampur University. He devoted his time in teaching and research in different Govt. and Non-Govt. colleges of Odisha since 1978.He has completed many minor research projects and major research projects funded by the University Grants Commission. He has published many research articles and edited many souvenirs of State level and National-level seminars funded by U.G.C., OBA and other Agencies. He is recipient of Prof. Amulya Kumar Panda Royal Teacher Award 2011 instituted by Royal College of Science and Technology, Bhubaneswar. He has edited a book on "Application Of Biology for Self employment" and co-authored a Fishery book entitled "Employment Through Aquaculture", "Biology of Wild Animals", "Wild Animals of India" and "Earning Animals" meant for students of undergraduate and post-graduate level. He has also edited a book entitled "Water for Survival". He is recipient of Emeritus fellowship by the University Grants Commission.